Test Engineering

WILEY SERIES IN QUALITY AND RELIABILITY ENGINEERING

Editor
Patrick D.T. O'Connor

Electronic Component Reliability:
Fundamentals, Modelling, Evaluation and Assurance
Finn Jensen

Integrated Circuit Failure Analysis:
A Guide to Preparation Techniques
Friedrich Beck

Measurement and Calibration Requirements
for Quality Assurance to ISO 9000
Alan S. Morris

Accelerated Reliability Engineering:
HALT and HASS
Gregg K. Hobbs

Test Engineering
Patrick D. T. O'Connor

Test Engineering

A Concise Guide to Cost-effective Design, Development and Manufacture

Patrick O'Connor

Stevenage, UK

JOHN WILEY & SONS, LTD
Chichester • New York • Weinheim • Brisbane • Singapore • Toronto

Other Wiley Editorial Offices

John Wiley & Sons Inc., 111 River Street, Hoboken, NJ 07030, USA

Jossey-Bass, 989 Market Street, San Francisco, CA 94103-1741, USA

Wiley-VCH Verlag GmbH, Boschstr. 12, D-69469 Weinheim, Germany

John Wiley & Sons Australia Ltd, 33 Park Road, Milton, Queensland 4064, Australia

John Wiley & Sons (Asia) Pte Ltd, 2 Clementi Loop #02-01, Jin Xing Distripark, Singapore 129809

John Wiley & Sons (Canada) Ltd, 22 Worcester Road, Etobicoke, Ontario M9W 1L1

Wiley also publishes its books in a variety of electronic formats. Some content that appears in print
may not be available in electronic books.

Library of Congress Cataloging-in-Publication Data

O'Connor, Patrick, D. T.
    Test Engineering : a concise guide to cost-effective design, development, and
    manufacture / Patrick D. T. O'Connor.
      p. cm.
    Includes bibliographical references and index.
    ISBN 0-471-49882-3
    1. Testing. 2. Engineering design. 3. Manufacturing processes. I. Title.

TA410.0175 2001
620′.0044 — dc 21   00–069308

British Library Cataloguing in Publication Data

A catalogue record for this book is available from the British Library

ISBN 0 471 49882 3 (Hbk)
ISBN 978 0 471 49882 7

Typeset in 10½/13pt Sabon by Techset Composition Ltd, Salisbury, Wiltshire

FSC
Mixed Sources
Product group from well-managed
forests and other controlled sources
Cert no. SGS-COC-2953
www.fsc.org
© 1996 Forest Stewardship Council

To Ina
For staying and supporting

Contents

7 Materials and Systems Testing   115

8 Testing Electronics   133

9   Software   149

Preface

Testing is usually the most expensive, time-consuming and difficult activity during the development of engineering products and systems. Development testing must be performed to ensure that designs meet requirements for performance, safety, durability, reliability, statutory aspects, etc. Most development testing involves integrated and multidisciplinary thinking and teamwork, between engineers proficient in the main technology areas and from design, development, manufacturing and maintenance specialisations. Most manufactured items must also be tested to ensure that they are correctly made, and many must also be tested in service. However, much of the testing that is performed in industry is based upon traditions, standards and procedures that do not provide the optimum balance of assurance versus cost and time. Misperceptions are commonplace, particularly the ideas that tests should not stress products in excess of their operating levels, that failures on test represent failures of the test, and that the purpose of reliability testing should be the generation of reliability statistics.

Testing is not just an engineering issue. Because of the importance and magnitude of the economic and business aspects, testing is an issue for management. There is often pressure to reduce testing because of the high costs involved, without appreciation of the effects on performance, reliability, safety and the potentially much higher costs of failures in service.

The main reason for insufficient or inappropriate testing seems to be that engineers have not developed a consistent philosophy and methodology for this essential activity. Testing is not taught as part of most engineering curricula, and there are no books on the wider aspects of the subject. Specialist areas are taught, for example fatigue testing to mechanical engineers and digital circuit testing to electronics engineers. However, a wide range is untaught, particularly multi-disciplinary and systems aspects. Engineering training tends to emphasise design. Testing and manufacturing are topics that attract less attention, and do not have the "glamour" of research and design. This is reflected in the generally lower esteem, status and salaries of engineers working in test and manufacturing. In some countries the near-disappearance of technician and apprentice training as routes to recognised engineering qualification has greatly reinforced this unfortunate trend. Engineering industry suffers shortages of talented engineers in these key areas. As a result, designs are often inadequately tested in development and products are inadequately tested in manufacture and maintenance. This creates high costs throughout the product cycle, damages competitiveness, and can lead to hazards.

Modern computer-based techniques, such as mathematical modelling, finite element analysis and electronic circuit design automation enable designs to be synthesised and analysed, so that most aspects of performance can be "tested" by the software. These powerful tools can greatly assist the creation of correct designs, and they can reduce the need for tests of the real product, and hence costs in development. However, they cannot reproduce the total reality of application conditions, such as environmental and manufacturing variations and interactions. Therefore their use and the results obtained should be treated as integral parts of the development test programme. This integration is seldom adequately taught or fully achieved.

Other adverse results of the lack of training on testing include the creation of standards that promulgate inappropriate and ineffective methods, and industry adherence to them.

A contributory reason why testing might be considered by many engineers, managers and users of modern products to be unimportant is the very high reliability achieved by most mass-produced products. How many of us have observed a failure of a microprocessor, a memory chip, a mobile phone, or a car engine? Modern engineered products are complex, inexpensive yet amazingly reliable, as a result of advances in technologies such as microelectronics, machining, surface metallurgy, plastics, lubrication, software, etc. However, we skate on thin ice: it is all too easy for failures to result from oversights or inadequacies in testing, and we can all remember cases, as engineers and as consumers. The element of uncertainty or chance looms large ("good luck" is rare in engineering: varying degrees of bad luck are the norm).

Testing should be based on a philosophy that would provide a foundation for all plans, methods and decisions related to testing of engineered products and systems. This is not to say that there will be one "right" way to test in any given circumstances, and that other ways are "wrong". Engineering in general, and testing in particular, is so much more difficult and includes so many more uncertainties than any other field of human endeavour, that there will always be scope for judgements, and even disagreements, on what and how to test. The design engineer might feel that there is no need for extensive testing, but the project engineer might think otherwise. The finance people will want to minimise the cost, and therefore the amount of testing. And so on. However, if all can agree on a philosophy that takes account of what is known, what is uncertain, and the other essential factors such as costs, timing, markets, regulations, safety, etc., then at least there will be a foundation for rational planning and decisions. *Intrinsic to this philosophy should be the maxim that effective testing should be considered as value-adding processes, not as costs.*

The objective of writing this book is to provide a text which can be used as the basis for teaching the principles of testing to all engineering students, as part of degree courses, and to provide a reference source for practising engineers and engineering managers. I hope that it will lead to a wider appreciation of good testing principles, and of the pitfalls of inappropriate methods.

A book such as this will inevitably be incomplete and will not reflect technology developments that occur after publication. I will attempt to maintain an updating system on my homepage (details below). This also includes listings of societies, software, equipment and services appropriate to design analysis and test. Suggestions and offers for contributions will be welcome.

I acknowledge with gratitude the considerable and generous help that I have received from specialists. In particular, Gregg Hobbs provided the inspiration and help on the sections on HALT and HASS testing, and for other sections. Ing. Rafat Malik of MSC Software provided the FE analysis in Chapter 5. Staff at nCode International provided information and inputs on fatigue analysis methods and software. Richard Baker of LDS gave valuable assistance on the subject of vibration testing. Jon Turino and Ben Bennetts contributed to the topic of electronics test. John Musa, Sam Keene and Sean O'Connor helped me on software testing. Brendan Davis provided information and insights on manufacturing test economics, in particular the concept of value-added test. Jim Morrison reviewed and improved the chapter on variation. Frank Everest and Sidney Dunn kindly reviewed the whole book and made many suggestions for improvement. Several companies provided figures and other information, and their contributions are acknowledged where appropriate.

Nevertheless, I am acutely aware that a book such as this will inevitably suffer from many imperfections, errors and omissions. I cannot improve upon the elegant expression of such shortcomings as was written by Dr. Peter Mark Roget in the preface to the first (1852) edition of his Thesaurus:

> Notwithstanding all the pains I have bestowed on its execution, I am fully aware of its numerous deficiencies and imperfections, and of its falling far short of the degree of excellence that might be attained. But, in a work of this nature, where perfection is placed at so great a distance, I have thought it best to limit my ambition to that moderate share of merit which it may claim in its present form; trusting to the indulgence of those for whose benefit it is intended, and to the candour of critics who, while they might find it easy to detect faults, can at the same time duly appreciate difficulties

I submit my book as a prototype, to be tested by its readers and users, using the principles described within.

Patrick O'Connor

pat@pat-oconnor.co.uk
Homepage (main): http://www.pat-oconnor.co.uk
Homepage (book): http://www.pat-oconnor.co.uk/testengineering.htm

February 2001

Series Foreword

Modern engineering products, from individual components to large systems, must be designed and manufactured to be reliable in use. The manufacturing processes must be performed correctly, and with the minimum of variation. All of these aspects impact upon the costs of design, development, manufacture and use, or, as they are often called, the product's life cycle costs. The challenge of modern competitive engineering is to ensure that life cycle costs are minimized, whilst achieving requirements for performance and time to market.

If the market for the product is competitive, improved quality and reliability can generate very strong competitive advantages. We have seen the results of this in the way that many products, particularly Japanese card, machine tools, earthmoving equipment, electronic components and consumer electronic products have won dominant positions in world markets in the last 30 to 40 years. Their domination has been largely the result of the teaching of the late W. Edwards Deming, who taught the fundamental connections between quality, productivity and competitiveness. Today this message is well understood by nearly all engineering companies that face the new competition, and those that do not understand lose position or fail

Concurrently with the philosophy and methods that took root initially in Japan and then spread back to the West where most originated, methods were developed in the USA to address the problems of unsatisfactory quality and reliability of military equipment. These included formal systems for quality and reliability management (MIL-Q-9858 and MIL-STD-758) and methods for predicting and measuring reliability (MIL-STD-721, MIL-HDBK-217, MIL-STD-781), MIL-Q-9858 was the model for the international standard on quality systems (ISO 9000), and the methods for quantifying reliability have been similarly developed and applied to other types of product.

The methods developed in the West were driven to a large extent by the customers, particularly the military. They reacted to perceived low acheivement by the imposition of standards and procedures, whilst their suppliers saw little motivation to improve, since they were paid for spares and repairs. By contrast, the Japanese quality movement was led by industry, who learned how quality provided the key to greatly increased productivity and competitiveness.

These two streams of development epitomize the difference betwen the deductive mentality applied by the Japanese to industry in general, and to engineering in particular, in contrast to the more inductive Western approach. The deductive approach seeks to generate continuous improvements across a broad front, and new ideas are subjected to careful evaluation. The inductive approach leads to inventions and 'break-throughs', and to greater reliance on 'systems' for control of people and processes. The deductive approach allows a clearer view, particularly in discriminating between sense and nonsense. However, it is not conducive to the development of radical new ideas. Obviously these traits are not exclusive, and most engineering work involved elements of both. However, the overall tendency of Japanese thinking shows up in their enthusiasm and success in industrial teamwork and in the way that they have adopted the philosophies of Western teachers such as Deming and Drucker, whilst their Western competitors have found it more difficult to break away from the mould of 'scientific' management, with its reliance on systems and more rigid organizations and procedures.

Unfortunately, the development of quality and reliability engineering has been afflicted with more nonsense than any other branch of engineering. This has been the result of the development of methods and systems for analysis and control that contravene the deductive logic that quality and reliability are acheived by knowledge, attention to detail, and continuous improvement on the part of the people involved. Of course Western minds have also made great positive contributions: we need only recall Shewhart's invention of statistical process control and Fisher's invention of statistical experiments, and of course Deming was an American. Therefore it can be difficult for students, teachers, engineers and managers to discriminate effectively, and many have been led down wrong paths.

In this series we will attempt to provide a balanced and practical source covering all aspects of quality and reliability engineering and management, related to present and future conditions, and to the range of new scientific and engineering developments that will shape future products. I hope that the series will make a positive contribution to the teaching and practice of engineering.

Patrick D.T. O'Connor
August 1994

1

Introduction

1.1 WHY TEST?

It is necessary to test engineering products for a number of reasons. These can be summarised under the following headings.

Design uncertainty

We must confirm that the design meets the specified requirements of performance, safety, reliability and durability. If we have complete confidence that the design will achieve all of the requirements, then performing tests to demonstrate this is, in principle anyway, unnecessary. However, the need for testing during development is a direct reflection of the amount of uncertainty in the design. In practically all engineering designs there will be an 'uncertainty gap' between what we think we know about the design and its reality in terms of all of the requirements that it is to achieve. For some new products there is little or no uncertainty, and so testing can be minimal. For example, a purely static simple item, such as a mounting bracket for a stationary component, can be designed with confidence that all features that affect performance and durability are understood and taken into account, since the task is simple in engineering terms. If, however, the bracket is to be used to support a component subjected to vibration, and it must also be as light as possible, then uncertainties begin to grow. If the consequences of failure are severe, it becomes even more prudent to perform tests to confirm the design calculations and analyses. Additional risks are incurred if the product is to be made in quantity, since the manufacturing processes will introduce variation, and therefore further uncertainty.

Of course, few engineering products are as simple as a mounting bracket. Even quite simple designs, such as a door actuator or an electronic timing circuit, will contain features that should be confirmed by testing. The amount of testing will obviously depend upon the designers' familiarity with the problems, and their knowledge, experience and skill. The formal processes of design, such as calculations, information on components and materials and analysis methods, often do not include aspects such as interactions, resistance to change or

1

degradation, and reliability. The effects of these on a system can be quite unpredictable, and ascertainable only by test. If the designs are shown by tests to be correct, then no redesign will be necessary, otherwise changes will have to be made and the changed design re-tested. Immediately we can see the strong connection between the experience, knowledge and skill of the design team and the cost of development.

Most engineering products are more complex than the examples given above. In the majority of cases it is not practicable for the design team to possess such knowledge that all problems can be solved, and all features optimised, without testing, analysing the results and making changes. In fact the process is often iterative over several cycles, especially for difficult designs. A new design of photocopier is a good example. Such a product contains a large variety of electronic circuitry (power, analogue, digital, high voltage), sensors, mechanisms, drives, controls, displays, etc. To be competitive it must have attractive price and performance features, and it must be reliable. It must also comply with standards for safety and electromagnetic compatibility. It is inconceivable that a design team, however skilled and experienced, could create a design for a new copier which would meet all of the performance, safety and reliability requirements without testing to find the shortcomings and to confirm the expected performance. For such a product the test programme will have to be long, intensive and expensive. There will be surprises and disappointments, and the tests will seldom run to the initial plan or budget. The reason for this is the uncertainty involved: we cannot know what failures will occur, or how often, or how much effort and time will be needed to correct them. The ultimate truth lies within the product itself, and testing is the only way to determine all of it.

Development test principles and methods are described in Chapters 6–9.

Manufacturing

When designs have been finalised and production is commenced, we must test the manufactured items to ensure that they have been correctly made and that they will comply with all of the requirements. Depending upon the type of product and its complexity, we might decide to test only samples, rather than all items made. Sometimes inspections or measurements may be used to supplement or replace tests, and in fact in most production situations there will be a mix of inspection, measurement and test operations.

Manufacturing testing will be covered in Chapter 10.

Variation

A very important reason for testing, during development and in production, is to minimise the effects of variation. All parameters of engineering products and all of the environmental conditions that they must endure are variable. In manufactured products, parameters such as machined dimensions, material properties, weld strengths, electronic component parameter values, etc., differ from item to item. Conditions of use, such as temperature, vibration stresses, humidity, electrical supplies, lubricant conditions, etc., also vary from application to application and over time. Variation is important to the extent that it can affect performance. Variations can also interact to generate combined effects that are greater than those of single variables. Engineering designs must take account of all of the important variables and their interactions, and the test programme, in development and in production, must ensure that the designs and the process controls are adequate in these respects.

Variation in engineering is the subject of Chapter 4.

Maintenance

Many engineering products require maintenance while in service. Maintenance tasks might involve routine checks and other tasks such as lubrication and calibration, as well as repairs and adjustments. It is often necessary to test the items after maintenance, to ensure that they function correctly and that repairs have been performed properly. Maintenance aspects of testing will be covered in Chapter 11.

Regulations

It is often necessary to test new products to demonstrate that they comply with regulatory requirements, such as for safety or compatibility. For example, pressure vessels must be tested in accordance with US ASME codes or British Standard specifications, and electronic equipment must be tested to demonstrate compliance with the regulations on electromagnetic compatibility. We will describe these types of tests in Chapter 13.

Contracts

Some projects are performed in accordance with contracts placed by the customer. Most military equipment and systems are governed by contracts, and these usually stipulate requirements for test. The expression *test and evaluation* (T&E) is sometimes used to encompass the total development test programme, including customer trials, for military systems.

1.1.1   Causes of failure to achieve design requirements

There are many reasons and causes why a product might fail to achieve the requirements set for the design (the *specification*). *Knowing, as far as is practicable, the potential causes of failure is fundamental to determining and executing rational and effective testing.* It is rarely practicable to anticipate all of the causes, so it is also necessary to take account of the uncertainty involved. Testing during development and in manufacture should address all of the anticipated and possibly unanticipated causes of failure, to ensure that their occurrence is prevented or minimised.

The main categories into which failures to achieve design requirements can be placed are as follows.

1. The design might be *inherently incapable*. It might be too weak, consume too much power, suffer resonance at the wrong frequency, etc. The list of possible reasons is endless, and every design problem presents the potential for errors, omissions and oversights. The more complex the design or difficult the problems to be overcome, the greater is this potential.

2. The item might be *overstressed* in some way. If the stress applied exceeds the strength then failure will occur. An electronic component will fail if the applied electrical stress (voltage, current) exceeds the ability to withstand it, and a mechanical strut will buckle if the compression stress applied exceeds the buckling strength. Overstress failures such as these do happen, but fortunately not very often, since designers provide margins of safety. Electronic component specifications state the maximum rated conditions of application, and circuit designers take care that these rated values are not exceeded in service. In most cases they will in fact do what they can to ensure that the in-service worst-case stresses remain below the rated stress values: this is called 'de-rating'. Mechanical designers work in the same way: they know the properties of the materials being used (e.g. ultimate tensile strength) and they ensure that there is an adequate margin between the strength of the component and the maximum applied stress. However, it might not be possible to provide protection against every possible stress application. If a 110 V appliance is connected to a 240 V supply it will probably fail, and a tyre will burst if sufficiently over-inflated.

3. Failures might be caused by *variation*. In the situations described above the values of strength and load are fixed and known. However, in most cases there will be some uncertainty about both. The actual strength values of any population of components will vary: there will be some that are relatively strong, others that are relatively weak, but most will be of nearly average strength. Also, the stresses applied will be variable. Figure 1.1 shows this more general load–strength situation. As before, failure will not occur so long as the applied load does not exceed the strength. However, if there is an overlap between the distributions of load and strength, and a load value in the high tail of the load distribution is applied to an item in the weak tail of the strength distribution so that there is overlap, then failure will occur. In cases where such overlapping distributions exist we say that there is *interference* between them. Obviously it becomes more difficult to design against failure in this kind of situation. We can no longer rely on simple, deterministic, values of load and strength. We now need to know how these values vary, particularly in the high tail of the load distribution and in the low tail of

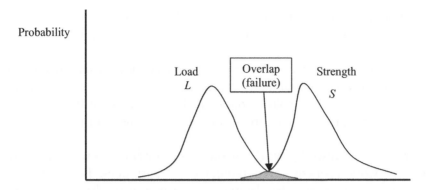

Figure 1.1   Load–strength variation

the strength distribution. Values near the average, or in the opposite tails, are of little interest to designers when considering this kind of failure. We will discuss the nature and effects of variation in engineering in more detail in Chapter 4.

4. Failures can be caused by *wearout*. We will use this term to include any mechanism or process that causes an item that is sufficiently strong at the start of its life to become weaker with age. Well-known examples of such processes are material fatigue, wear between surfaces in moving contact, corrosion, insulation deterioration, and the wearout mechanisms of light bulbs and fluorescent tubes. Figure 1.2 illustrates this kind of situation. Initially the strength is adequate to withstand the applied loads, but as weakening occurs over time the strength decreases. In every case the average value falls and the

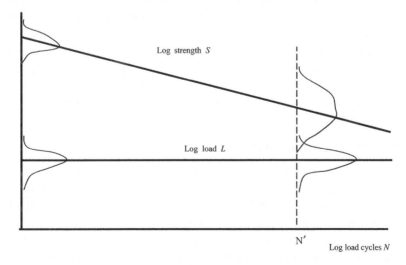

Figure 1.2   Time-dependent load–strength variation

spread of the strength distribution widens. This is a major reason why it is so difficult to provide accurate predictions of the lives of such items.

5. Failures can be caused by other time-dependent mechanisms. Battery run-down, creep caused by simultaneous high temperature and tensile stress, as in turbine discs and fine solder joints, and progressive drift of electronic component parameter values are examples of such mechanisms.

6. Failures can be caused by *sneaks*. A sneak is a condition in which the system does not work properly, even though every part does. For example, an electronic system might be designed in such a way that under certain conditions incorrect operation occurs. The fatal fire in the Apollo spacecraft crew capsule was caused in this way: the circuit design ensured that an electrical short circuit would occur when a particular sequence was performed by the crew. Sneaks can also occur in software designs.

7. Failures can be caused by *errors*, such as incorrect specifications, designs or software coding, by faulty assembly or test, by inadequate or incorrect maintenance, or by incorrect use. The actual failure mechanisms that result might include most of the list above.

There are many other potential causes of failure. Gears might be noisy, oil seals might leak, display screens might flicker, operating instructions might be wrong or ambiguous, electronic systems might suffer from electromagnetic interference, etc. The important point to appreciate here is that failures are by no means always clear-cut and unambiguous. They are often open to interpretation and argument. They also differ in terms of importance (cost, safety, other effects). Therefore we must be careful not to apply conventional scientific, deterministic thinking to the interpretation of failures. For example, a mere count of total reported failures of a product is seldom useful or revealing. It tells us nothing about causes or consequences, and therefore nothing about how to improve the situation. This contrasts with a statement of a physical attribute such as weight or power consumption, which is usually unambiguous and complete.

1.1.2 Reliability and durability

Reliability is the ability of an item or of items to keep operating, or to be available for operation, over a period of time without failing. Reliability can be expressed as a mathematical probability of not failing or of being available for use. *Durability* is the ability of an item to withstand wearout mechanisms such as fatigue, wear, etc. Durability is usually expressed as a minimum time before the occurrence of wearout failures. The development test programme must include tests to provide assurance of reliability and durability.

Reliability and durability are covered in Chapter 4.

1.1.3   Environments

Most engineering products must operate in a range of environmental conditions. These can include temperature extremes and changes, vibration, shock, humidity, corrosive atmospheres, electrical power conditions, and many others. In particular cases other conditions might be important, such as altitude, solar or other radiation, acoustic noise, electromagnetic interference, etc. The environmental conditions can influence performance, reliability and durability, so their effects must be considered in planning the test programme. The extreme values of the environmental conditions, the times and cycles over which they occur, and their rates of change, must all be considered. Environmental effects often are interactive, so that the effects of combined environments and operating stresses can be more severe than any one effect in isolation. For example, the growth rate of fatigue cracks in materials such as steel can be accelerated by the presence of corrosion, and the quality of electronic soldering processes can be affected by combined factors such as solder temperature, composition, preheat, and several others. Therefore the test programme should cover the real, combined, environmental conditions as far as is practicable.

The human aspects of the environment are also important, since people, working as producers, packers, shippers, installers, operators and maintainers, all influence the quality and life of the product. Therefore the test programme should also include the human aspects, not only physical operating and environmental conditions.

1.2   HOW TO TEST?

1.2.1   Test to succeed, or test to fail?

Product testing can be considered to be in two categories. In the first, we hope that the product will not fail. A test to demonstrate that a design will comply with the specification, for example that it will not consume more than the power specified, or that it will comply with a mandatory safety standard, should be planned on the expectation of success.

On the other hand, tests to provide assurance of strength, reliability or durability of a design must be planned to generate failures, or at least evidence that failure will not occur, or will be very unlikely to occur, within the product's expected conditions of use and lifetime. The uncertainty and variability inherent in forecasts of such properties mean that testing to generate failures outside the operating regime is usually the only practicable way of providing assurance that failures will not occur inside. Therefore the test programme must distinguish between the tests that should demonstrate success, and those that are planned to generate failures.

Production tests must not damage good items. However, they should be able to show if items are not good (weak, defective, etc.), and the most appropriate way to do this, in most cases, is by causing them to fail.

It is not necessarily the case that the whole product must be tested to failure. At one extreme, a civil engineering structure such as a bridge cannot economically be tested to failure in fatigue due to cyclic loading. Therefore the bridge designer uses the knowledge of tests to failure of samples of the bridge steel and includes strength margins that guarantee that such failures cannot occur. On the other hand, the designer of an aircraft wing or pressure cabin is denied the luxury of such large safety margins, and so must create a compromise between weight and safe life as well as other features, such as ease of crack detection, provision of crack stoppers, and minimisation of stress concentrations. Such a compromise must be tested to failure to provide assurance that the inevitably uncertain analyses are sufficiently conservative.

1.2.2 Accelerated test

Since both test time and units to test are major cost considerations in almost any development programme, it is essential that the right balance is struck between investment in testing and the payback. One way of economising on test time is to apply *accelerated testing*. This involves the use of stress conditions, including cycling frequencies where appropriate, higher than those expected to be encountered in service. One form of accelerated testing is *step-stress* testing, in which the stress levels are progressively increased until failure occurs. Accelerated tests can greatly increase the cost effectiveness of a test programme. However, it is necessary to investigate failures carefully to ensure that they are relevant. The criterion of relevance must be the possibility of occurrence in service, and the evidence they present of design safety margins. In situations where the relationship between stress intensity and failure is fairly well known, for example fatigue, the data and models can be used to evaluate the reliability at lower stress levels. However, such relationships are subject to considerable imprecision and should be used cautiously.

Design engineers often are reluctant to agree to testing of early prototypes at conditions that exceed those specified, and they might consider failures of such early models generated under such conditions to be unrepresentative, and therefore not worth investigating or taking action to prevent. However, it is important to investigate every failure that occurs, under any conditions. The criterion for relevance and action must be whether the failure could occur on production items in use. Therefore disciplined, 100 percent failure reporting and investigation must be imposed from the start of the test programme.

Accelerated tests are also appropriate for production testing, to discover product weaknesses by causing failures and to reduce test time.

Accelerated testing principles and methods are described in Chapter 6 and (for manufacturing) in Chapter 10.

1.2.3   Testing systems and components

When a product consists of several subsystems or lower-level components, it is necessary to consider testing at different levels of assembly. Testing of lower-level components and sub-assemblies is less expensive and can be more effective in exploring the limitations of the designs than testing at the level of the complete system. For example, the performance and reliability of an electronic engine fuel control unit can be explored using several units, operated over a wide range of conditions, far more cheaply and effectively than by reliance on testing of the vehicles in which the units will operate. Lower-level tests can also be carried out earlier than system-level tests. Of course, most systems comprise a range of components and sub-assemblies, ranging from simple to complex, from mature to innovative, and from standard to unique to the new system being developed. Items that are unique, innovative or complex must be given priority, since they usually represent the greatest risks. When items are bought from external suppliers, the system developers must ensure that the tests applied are adequate, either by reviewing the suppliers' test programme and results, by conducting separate tests, or a combination of both.

Component and sub-assembly testing provides assurance that the products concerned can operate under the conditions expected. However, the test conditions can differ in important and subtle ways from the actual conditions in the system. For example, the way that a sub-assembly is fitted into a system might induce static and dynamic mechanical forces that are different from those applied in tests of the sub-assembly. Equally important, system-level interactions might occur that are not simulated in lower-level tests. Typical of these are the effects of location and connections on electromagnetic compatibility and timing of electronic units, and the interaction of engine and transmission dynamics on vehicle noise and vibration.

Finally, it is not normally possible to confirm the performance of a system purely by synthesising the performance of its components and sub-assemblies, due to the many uncertainties and interactions that exist. Therefore it is essential that the complete system is tested, as well as its components and sub-assemblies.

1.2.4   Testing the technologies

The three major technology branches that are involved in most engineering are mechanical (including materials), electronics and software. Most modern products and systems utilise some combination of these. All electronics products involve mechanical design. Component packaging, electronic connections, and mechanical stresses due to vibration, shock and temperature change are some of the important aspects of mechanical engineering design of electronics systems that must be taken into account, and covered by the test programme. Many electronics designs also involve software as part of the operating system, and also for testing during manufacture and maintenance. The mechanical/materials aspects of engineering products can lead to overstress or wearout types of failures, as described earlier. Electronic components and circuits can suffer from these, as well as from purely electrical/electronics causes, such as electrical overstress, functional problems such as parametric, logical or timing errors, and electromagnetic interference. Software *per se* can fail only if there are logical errors in the design: there are no time-dependent processes of the types that can cause deterioration or other changes in materials, and also no variation.

These technology-related factors must be taken into account during testing, both in development and during production. Sometimes they can be treated as components or sub-systems, but they must also be included in testing of the overall system. This means that the engineers planning and running the tests must have sufficient knowledge of all of the relevant technologies being applied, and the test programme must cover all of the technology aspects in an integrated way. We will discuss the three main technology aspects of test in Chapters 7, 8 and 9.

1.2.5   Testing the processes

For any product that is to be manufactured in quantity it is imperative that the test programme includes *testing of the manufacturing processes*, not only of the product design. The more that the costs of the manufacturing processes influence the product costs, the more important it is that the development test programme includes testing of processes. Manufacturing process costs depend upon the difficulty or novelty of the processes, the number of processes, and the cost of the production facilities. The cost of facilities (jigs, automation, test equipment, etc.) will depend upon the product and on the quantity to be produced. Manufacturing costs are also influenced by the costs of failures: diagnosis, repair, scrap, etc., so the quality of the processes (assembly, inspection, test, etc.) also needs to be addressed.

Normally, the larger the quantity to be produced the more need there is to invest effort and resources to minimise the marginal cost of production by the use of capital facilities, since the fixed costs can be amortised over larger production quantities. However, when such facilities and the methods by which they will be used are determined as part of the overall design and development programme, then their effectiveness must be tested. Unfortunately it is not uncommon for products to be designed and developed, to the point at which drawings and process instructions are released for manufacture, and then the production people are expected to begin manufacture to schedules and budgets determined during design. Of course such schedules and budgets assume that there will be no problems, even if they do allow for initial learning by the production people. However, process problems always occur. The only way of avoiding them during the manufacturing phase is to test the production processes during development so that they can be improved and refined before production commences.

Problems can be removed by changing the process, or by changing the design, or possibly both together.

As with product testing, process testing should begin as early as possible to give time for improvements to be developed. As far as practicable the processes should be performed by the people who will operate them, using the equipment (tools, jigs, test equipment, etc.) that will be used in production. If the processes are performed by development engineers, using unrepresentative equipment, it is very likely that real process problems will not be detected. The production people should be integrated into the development test programme at a sufficiently early stage, and their findings and recommendations must be given as much attention as those related to product design aspects.

Maintenance, including servicing and repair, of the product in service also involves processes, many of which are the same as or similar to the manufacturing processes. Therefore the product support people should be involved in the development test programme, and maintenance operations must be tested and refined.

Integrating manufacturing and maintenance process testing into the design and development test effort can generate considerable benefit when the product enters the production and use phases. Manufacturing and maintenance people can contribute useful inputs, based on experience, to help to avoid later problems. The initial cost per unit produced will be reduced, and the subsequent cost reduction rate due to learning effects will be increased, since fewer problems will be encountered and there will be a core of trained, experienced production people at the start. The same applies to the maintenance aspect, as the service people will also have benefited from the experience gained during development testing. The experienced production and service people will be better equipped to train others. Also, the need for costly changes to designs and processes, particularly changes that delay production, will be greatly reduced.

1.3   ANALYSIS AND SIMULATION

All engineering designs are analysed to ensure, as far as is practicable, that they will achieve their requirements. Analysis methods cover the range of parameters, application conditions, etc. relevant to the design, so the ability to withstand stresses, operate for specified times, and other criteria will be determined during the design phase. Other methods of analysis can be applied, to determine aspects such as the responses to failures of components or to variations of parameters. All analyses can be considered to be 'tests' of the design, to confirm that it meets the requirements and to help to identify aspects which should be tested more directly.

Computer simulations are well developed and are widely used for analysing system designs. Software based on finite-element analysis (FEA) can be used to determine distributions of mechanical stress, simulate structural behaviour under stress or vibration, and determine temperature and electromagnetic field distributions. Computational fluid dynamics (CFD) software can be used to model gas and liquid flow behaviour. Two- and three-dimensional drawing software can be used to analyse designs for tolerances, conflict of moving parts, ease of assembly, and accessibility. Using capabilities such as these to test the performance of designs is usually less expensive than making and testing hardware. The testing can also be started earlier.

Virtual reality (VR) software enables even greater analytical capability in some design situations. Engineers can use it to explore aspects such as man–machine interactions and assembly and maintenance operations.

The test programme should be based upon the results of design analyses and simulations, and the analysis results should be used to complement the data derived from tests. Analysis and simulation can in some cases replace or reduce the need for testing. However, it is very important that any limitations of the analyses and simulations are appreciated and understood. Analyses and simulations can mislead. The product tells the truth.

Analysis and simulation methods, as they relate to testing, are described in Chapter 5.

1.4   GOOD AND BAD TESTING

For nearly every kind of testing that is performed, there are examples of excellent practice and of practice that is sub-optimal, sometimes downright counter-productive. Some examples of the latter from the author's experience are as follows.

- A project director, managing the development of a military system that involved novel technologies and high risk of failure, stated that there would be no environmental testing because 'our engineers are paid to get their designs right'.

- Railway systems, particularly new locomotives and trains, are subjected to minimal development testing in comparison with systems of equivalent complexity and risk in other industries, such as cars, aircraft and military systems. The reasons are not based upon any logic, but entirely on tradition. For most of its history, rail vehicle engineering has consisted of relatively proven technology, applied by a small number of famous designers. Also, there was nowhere to test a new train except on the rails with all the other traffic. This limited testing tradition suddenly came unstuck from about the 1980s when rail vehicle design belatedly but rapidly included a range of advanced technologies, such as AC electric traction, air conditioning, digital system controls, passenger information systems, etc. When new British trains suffered many failures of traction motors during a spell of unusually fine, dry snowfall, due to icing of the cooling air ducts, a rail spokesman famously stated that 'it was the wrong kind of snow'. European rail vehicle suppliers are now building test tracks and environmental test facilities. By contrast, aircraft are subjected to intensive flight testing, and a similar icing problem on the engine intake ducts of the Britannia airliner was discovered and corrected before entry to service.

- A large diesel engine was selected to power a new diesel-electric freight locomotive. The engine was a well-proven machine, with previous applications in ship propulsion, power generation, etc. To provide assurance for the rail application, one engine was subjected to a 'standard type test', involving 150 hours' continuous running at maximum rated power. It passed the test. However, in rail service it proved to be very unreliable, suffering from severe cracking of the cylinder block. The problem was that the previous experience involved long-duration running under steady and mostly low-stress conditions, which are totally different from the very variable, often high-stress, rail application. Also, in the previous applications, and in the 'type test', the coolant supply was large enough to ensure that it was always cool on entry to the engine, but the locomotive coolant tank was much smaller, so that the inlet temperature was very variable. The combination of variable duty cycles and variable coolant temperature led to early fatigue-induced cracking of the block.

- A contract for the development of a complex new military system included a requirement that the reliability be demonstrated in a formal test before the design could be approved for production. The test criterion was that no more than 26 failures were to occur in 500 test hours, in the specified test conditions. When questioned, the customer 'reliability expert' accepted that the test criterion would not be achieved if 27 minor failures occurred, but would be achieved with 25 major failures. (This is one example of many such cases: we will discuss 'reliability demonstration' testing later.)

- A new airline passenger entertainment system was developed and sold, for installation in a fleet of new aircraft. Reliability was considered to be a critical performance requirement, since failure of any seat installations would lead to passenger complaints. A test programme was implemented. However, cost and time constraints resulted in inadequate testing and problems detected not being corrected. Reliability was so poor when the system was installed and put into service that it eventually had to be removed, and the project terminated.

- A manufacturer of electronics systems submitted samples of all production batches to long-duration test at 40°C. When asked why, the Quality Manager replied 'to measure the reliability'. Further questioning revealed that the systems had been in worldwide service for some time, that in-service reliability performance was well known from maintenance and utilisation records, and that the testing had not generated any problems that were not already known. So I said: 'but you know the reliability, this test is very expensive and delays delivery, so why not stop doing it?'. Months later they were still doing it. Apparently the reason was that their written company procedures said it had to be done, and changing the procedures was too difficult! (We will discuss more situations like this later.)

- The US Military Standard for testing microcircuits (MIL-STD-883) requires that components must be tested at high temperature (125°C) for 168 hours. This requirement was later copied into other national and international standards. The reason for choosing this unnecessarily long and expensive test time? There are 168 hours in a week!

- Among the sparse examples of recent books on aspects of testing, one makes no mention of accelerated tests and another actually condemns the idea of testing production electronics hardware at stresses higher than might be experienced in service.

- Some 'experts' argued that systems that relied on software for safety-critical functions such as aircraft flight controls could never be considered to be safe, because it is not possible to prove by mathematical analysis or testing that failures will *never* occur. (We cannot prove that for pilots or mechanical controls either, but software does not make the mistakes humans make, and mechanical controls do break.)

- A military system, in service with different forces, displayed much worse reliability in army than in air force use. Investigation of the causes revealed that the army was following its traditional procedure of testing the whole system every day, whilst the air force's procedure was to test only if problems were reported. (How often do you test your TV?)

- No one seems to be able to report a completed engineering development project where too much testing had been performed. Nearly all engineering projects could have benefited from more testing, and wiser testing.

The common reason for such inconsistency and frequent poor practice is the fact that testing is very seldom taught in engineering curricula. Specialist technology aspects, such as materials test or digital circuit test methods, are usually included as appropriate, but university engineering education tends to emphasise theory and design. There have been no books on the wider aspects (philosophy, multidisciplinary aspects, economics, management), and nearly all books on engineering management ignore testing. Consequently, few engineers possess sufficient broad knowledge of the subject, and managers of engineering, many of whom are not professional engineers, are inadequately advised and perceive testing as expensive and problematic. It is, therefore, essential that it is managed effectively.

1.5 TEST ECONOMICS

1.5.1 Development

Because testing during development is perceived to be expensive, there will usually be pressure to reduce it, or at least not to extend it beyond what is considered 'acceptable'. On the other hand, the engineers involved might be concerned that, by the end of the test programme, there might still be features that could benefit from further testing and development, and failures might have occurred that cannot be guaranteed not to recur in service. There is no way of avoiding this dilemma for complex modern products such as domestic appliances, cars, electronic systems, trains and planes. However, the balance between the cost and effectiveness of the test programme in providing assurance and generating improvement can be optimised by concentrating on three principles of product design and development:

1. Appreciating that the long-term costs of failures to achieve design requirements and of product failures in service are nearly always far higher than the costs of well-managed design analysis and testing to show where improvements can be made. *Well-planned and executed analysis and testing are value-adding activities, not burdens.*

2. Ensuring that the engineers concerned have the knowledge, experience and leadership to create the maximum proportion of designs (product and process) that are correct.

3. Planning a test programme that maximises the chances of revealing all designs that are not.

   How can the development test programme be optimised? The major contributors to the cost of testing are:

- The articles being tested
- The manpower costs (people multiplied by time)
- Test facility costs, and for these time is also a multiplier
- The opportunity cost of the delay in starting sales
- The downstream opportunity costs of warranty, reputation, fixes, etc.

Therefore we must obtain the maximum information and improvement, using the minimum number of items and facilities, for the shortest practicable time.

The number of items to be tested will obviously depend on the cost of manufacture of prototypes. Each prototype aircraft or spacecraft represents a huge additional cost to the project, so the number to be provided for test must be minimised, and the test programme must be carefully planned to make the best use of these. For products such as engines, cars and electronic test equipment, the cost of prototypes is high, but not so high that the number of prototypes to be used in the test programme needs to be cut to a bare minimum, since the cost of insufficient testing, or late completion of testing, is likely to be much greater than the subsequent costs of undetected and uncorrected problems. This is particularly true if the product is to be mass produced, or sold in a very competitive market. For low-cost items, such as electronic or mechanical components, the test programme need not be constrained by the cost of providing the hardware.

If testing is delayed until the designs are considered to be complete, and nearly ready to be released for production, then problems discovered during testing at such a late stage could seriously disrupt the overall project programme. Therefore the correct approach in any development programme is to start testing as soon as hardware can be provided, even if it is only partly representative of the expected final design, in order to discover problems as early as possible. This early testing must then be followed by repeat tests of the improved designs to demonstrate the effectiveness of the improvements and to identify any further problems that might exist.

In many cases test costs and durations can be reduced while effectiveness is increased. The time and other resources needed for development testing can be greatly reduced by accelerated testing. Later chapters will describe how.

Development test economics are discussed in Chapter 14.

1.5.2 Manufacture

Testing as a part of manufacturing is also expensive. However, the concept of value-added testing applies to this phase also. We must first ensure that as few errors are made in the manufacturing processes as possible, and that inspection and test will detect those that are made. Testing must not be allowed to cost more than the costs of undetected failures (warranty, reputation, safety, etc.). However, such downstream costs can be very high, and are usually also very uncertain. Therefore optimising the manufacturing test programme is important, but difficult.

Manufacturing test economics are discussed in Chapter 10.

1.5.3 Maintenance

Many products and systems must be tested during their time in service, to ensure that they continue to operate properly and safely. The criterion of value should be applied to testing during the in-service phase, as well as to other maintenance tasks such as cleaning, inspection, repair, replacement, etc. We will discuss these aspects in Chapter 11.

1.6 MANAGING THE TEST PROGRAMME

Testing in development is the last chance to make sure that the design is correct and inherently reliable. Testing in manufacture is the last chance to ensure that the product is built correctly. Testing in service attempts to ensure that the product continues to operate correctly. The nature of the tests that should be applied is different for each phase.

Effective management of the test programme is as important as management of any of the other engineering disciplines and processes. It should be clear from the discussion so far that testing should not, however, be regarded as a topic that can be considered or managed in isolation from the other disciplines. The requirements for test, the methods to be applied and the economics and timing are all strongly influenced by factors that are the concerns of management. The most important of these are:

- The capabilities and experience of the design team in relation to the risks

- Regulatory and other requirements

- Markets and competition

- The product's environment

- Manufacturing methods and quantities

- Supplier selection and control.

Therefore testing must be managed as an integral part of the whole engineering process, in terms of the overall philosophical approach as well as on a project-by-project basis. It also follows that every designer of products and processes must understand the test philosophy and the methods that are appropriate, during development as well as in production and service. Likewise, manufacturing and product support engineers must also know what testing is performed during development, and must use the information to help optimise manufacturing and in-service tests. *All engineers must be taught testing.*

Test costs, particularly during development, arise early in the product cycle. They are usually quite easy to estimate and they are high. The costs of inadequate testing arise later. They cannot be estimated with anything like the same confidence, but they are even higher, and often very much higher. Testing is expensive but necessary, and in principle adds value to the product over the long term.

We will describe and discuss the management of testing in Chapter 14.

2

Stress, Strength and Failure of Materials

2.1 INTRODUCTION

Most engineering products must withstand stresses during their operating lives, and sometimes also during manufacture. Stresses can be mechanical, electrical, or caused by other physical processes. In principle, so long as the resisting strength of the item exceeds the stresses applied, failure will not occur. In order to prevent failures we need to understand and apply the physical laws and other factors that influence stresses and strengths. Our understanding of these derives from basic knowledge and experience, but in practical engineering this is often incomplete and uncertain. To ensure that we do obtain the full understanding necessary we must usually perform tests. Determining the most appropriate tests and interpreting the results depends upon sound appreciation of the science involved. In this chapter we will review the causes and effects of the stresses that can lead to failure, in sufficient detail to provide a basis for effective testing, particularly during development.

It is important to remember that the theoretical relationships that enable us to calculate stress and strength values imply a level of knowledge and precision that is seldom attainable in practice. As will be explained in Chapter 4, all parameters, applied stresses and environmental conditions are variables. They are not fixed, deterministic quantities, and the range and nature of the variations can often be wide and uncertain. This means that the interactions between them are even more uncertain. When time-dependent effects exist, such as fatigue, wear, corrosion or electrical parameter change, the uncertainties grow by further orders of magnitude. These uncertainties will be covered in relation to the specific stresses described later.

It is convenient to consider three aspects of failure. The underlying causes (physical, chemical, human) are the *failure mechanisms*. The *failure modes* are the immediate effects within a product or system, such as softening, open circuit, etc. The *failure effects* are the effects as observed by product or system behaviour, such as breaking, loss of output, etc.

Reference 1 describes failures in the context of reliability engineering.

19

2.2 MECHANICAL STRESS AND FRACTURE

Mechanical stress can be either *tensile, compressive* or *shear*. Tensile stress is caused when the material is pulled, so that the stress attempts to overcome the internal forces holding the material together. Typical material behaviour in tension is shown in Figure 2.1. This shows that, as stress increases, the material stretches proportionally to the stress (the *elastic region*), then begins to stretch more rapidly (the *plastic region*), and finally fractures. In the elastic region the material will return to the original unstressed length if the stress is removed. The amount of deformation is called the *strain*. In the plastic region the material will retain some or all of the deformation if the stress is removed. Fracture occurs when sufficient energy has been applied to overcome the internal forces.

Stress is the load per unit cross-sectional area, conventionally expressed as σ, and is measured in kg/m^2, lb/in^2 (psi), or pascals (Pa) (N/m^2). The strain (ε) is the ratio of the change in length to the original length. The relationship between stress and strain is described by *Hooke's Law*:

$$\sigma = E\varepsilon$$

where E is *Young's modulus*, or the *modulus of elasticity* for the material. A high value of E indicates that the material is *stiff*. A low value means that the material is soft or *ductile*.

The strength of a material in tension is measured by its *yield strength* (the stress at which irreversible plastic deformation begins) or *ultimate tensile strength* (UTS), the stress at which fracture occurs. Note that the UTS might be lower than the yield strength.

When a specimen is subjected to tensile stress, narrowing or 'necking' occurs, so that the cross-sectional area is reduced. This causes the 'true' stress level to

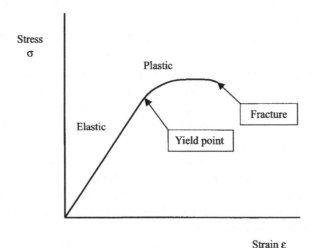

Figure 2.1   Material behaviour in tensile stress

increase compared with the 'engineering' stress calculated on the basis of the original unstressed cross-section. However, engineering design practice generally limits stress to not more than about 0.2 percent strain, so the engineering stress–strain relationships are mostly used.

The elastic/plastic/fracture behaviour of a material is determined by its atomic or molecular structure. Atoms in solids are bound together by the interatomic or intermolecular attractive forces. E is proportional to the interatomic spacing, and is reduced if temperature is increased. Elastic deformation extends the interatomic distances. Plastic deformation occurs when the energy applied is sufficient to cause the atomic planes, for example in a crystal, to slip along the lattice structure and take up new stable conditions. Fracture occurs when the stress exceeds the UTS.

Material surfaces contain energy, in the same way as the surface of a liquid possesses surface tension, due to the fact that the interatomic attractive forces between atoms at the surface can act in only two dimensions and so do not cancel as they do within the bulk material. In solids this energy is much higher than in liquids. When fracture occurs, two new surfaces are created. This extra energy is imparted by the applied stress which causes the fracture. Knowing the surface energy of a material enables us to determine the theoretical strength. This far exceeds what we actually measure, by factors of 1000 to over 10000. The reason for the difference is that some plastic deformation occurs at stresses much lower than the theoretical elastic limit, as actual materials contain defects that create stress concentrations, for example dislocations within crystal planes of crystalline materials (metals, metal alloys, silicon, carbon, etc.), and between molecular boundaries in amorphous materials like plastics. Very pure single crystals, such as carbon fibres, can be produced with strengths that approach the theoretical values. The practical strength of a material can be determined only by tests to failure, though theoretical knowledge of aspects such as crystal structure, uniformity, etc., enables materials scientists to make approximate forecasts of strength.

Another important material property is *toughness*. Toughness is the opposite of *brittleness*. It is the resistance to fracture, measured as the energy input per unit volume required to cause fracture. This is a combination of strength and ductility, which is represented by the area under the stress–strain curve. Figure 2.2 shows this schematically (and very generally) for different material types.

The different patterns of behaviour represent the properties of ductility, brittleness and toughness. A ductile, weak material like pure copper will exhibit considerable strain for a given stress, and will fracture at low stress. A tough material like Kevlar or titanium will have little strain and a high UTS. A brittle material like cast iron, glass or ceramic will show very little strain, but lower resistance to rapid stress application such as impact loads. Material properties, especially of metals, vary widely as a result of processes such as heat treatment and machining. In practice materials are applied so that the maximum stress is always well below the yield strength, by a factor of at least 2.

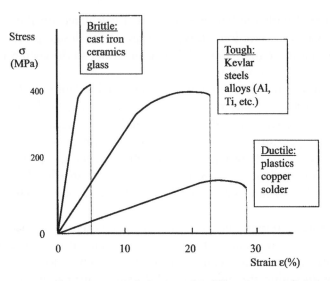

Figure 2.2 Tensile stress–strain behaviour for different materials (generalised)

A crack will grow if the energy at the crack tip is sufficient to overcome the interatomic forces, and thus open it further. *Griffith's Law* expresses this:

$$\sigma = \sqrt{(2E\gamma/\pi a)}$$

where:

σ = maximum stress at crack tip (note: not the average stress in the material)
E = modulus of elasticity
γ = surface energy
a = half crack length.

The maximum stress at the tip of a crack (or at any other defect or *stress raiser*) is proportional to the applied total stress, the size of the crack or defect, and the sharpness of the tip around which the stress is applied. The ratio of maximum stress to applied stress is the *stress concentration factor*. Whilst the total or average stress can usually be determined quite accurately, using methods such as finite element analysis, the maximum stress around a stress raiser, and thus the strength, is often much less certain. Stress concentrations can be reduced by designing to provide adequate radii of curvature on corners of stressed components, ensuring that material surfaces are smooth, and, in the case of cracks in sheet material, by drilling a hole at the tip of the crack to increase the radius.

Compressive strength is much more difficult to analyse and predict. It depends upon the mode of failure (usually buckling for most engineering materials and components such as steel or aluminium alloy vehicle panels, struts, and electrical connector pins) and the shape of the component. Compressive fracture can also occur, particularly in brittle materials.

Structures that have bending loads applied are subjected to both tensile and compressive stress. The upper part of a loaded cantilevered beam will be in tension, whilst the lower part will be in compression.

Stress can also be applied in *shear*. A common practical example is solder joints that connect surface-mounted electronic components (integrated circuit packages) to circuit boards: during operation the temperature rise in the component causes thermal expansion relative to the circuit board, thus applying shear stresses to the solder joints.

The discussion above has presented a very brief overview of the topic. In most cases of applied stress the material behaviour is more complex, since combined effects occur. For example, a component in tensile stress will be caused to be compressed in the directions perpendicular to the tensile stress, so there will be a compressive stress also. Bending loads cause varying tensile and compressive stress from top to bottom of the beam, and therefore shear stress within the beam. Fracture in compression might be caused by shear stresses generated in the material. For all of these reasons it is nearly always necessary to test structural components to determine their true strength, especially if the designs are not simple.

2.2.1 Fatigue

Mechanical stresses might be applied continuously, as in a suspension bridge cable, or cyclically, as in an aircraft wing in turbulent flight. The pattern of cyclic stress can, of course, vary from small changes over relatively long periods to high stress cycling at high frequencies, such as in a component subjected to vibration. The stresses could be caused by mechanical loads (vibration, vehicle suspension loads due to road and driving conditions, centrifugal forces on helicopter blade attachments, pressure loads in hydraulic or pneumatic components, etc.). Cyclic mechanical stress can also be caused by temperature changes, leading to differential thermal expansion. Such loads occur in engine components (turbine blades, pistons, cylinder blocks, etc.), in electronic component attachments, etc.

It is usually relatively easy to design products and processes so that the margin between static strength and static stress is adequate. However, cyclic mechanical stresses well below the yield or fracture stress can cause progressive weakening, so that the resisting strength is reduced, eventually resulting in failure. This mechanism is called *fatigue*. A component suffers from cumulative fatigue damage if subjected to cyclic, reversing stresses whose values exceed a level called the *fatigue limit* or the *endurance limit*. The mechanism of fatigue damage is the formation of microcracks, which grow as energy is applied to the crack tips due to the cyclic stress application. Initiation and growth rate of the cracks vary depending upon the material properties and surface and internal conditions. The material property that imparts resistance to fatigue damage is the toughness. As described above, the stresses around the tip of a crack or other defect (such as a machining scratch on a component or a void or inclusion in a casting or forging) are much higher than those in the bulk of the material, so concentrating the energy at these locations. We can demonstrate this easily by repeatedly bending a straightened paperclip through 180°. Being of ductile material, the clip will not fracture on the first bending. However, the alternating tensile and compressive stresses will generate cumulative fatigue damage, leading to fracture after typically about 20 cycles. If we now repeat the experiment, but now test paperclips which have been lightly cut with a sharp modelling knife, they will fracture in typically five cycles or fewer.

The number of cycles to failure (fracture) depends upon the amount by which the stresses exceed the fatigue limit, and the fatigue properties of the material. For most metals, the relationship known as *Palmgren–Miner's Law* expresses this:

$$\Sigma(n_i/N_i) = 1$$

where n_i is the number of cycles to failure at a particular stress as determined by tests, and N_i is the number of cycles applied at that stress in the application. For example, if we subject a specimen of the material to a defined stress cycling tensile test over a known stress range, and it failed after n cycles, we could use the data to determine the number of cycles to failure at some other pattern of stress applications in service.

The relationship can also be shown on the *S–N curve*, where S is the stress applied and N is the number of cycles to failure at that stress (Figure 2.3 is a typical example). The exact number of cycles to failure is, however, highly

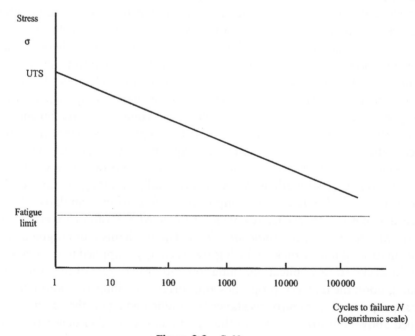

Figure 2.3   *S–N* curve

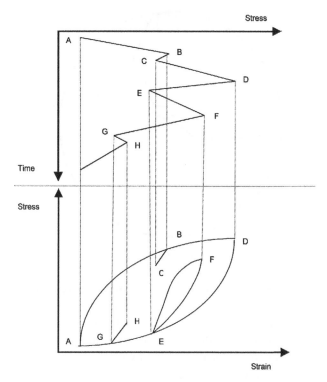

Figure 2.4   Rainflow plot

variable, even for nominally identical samples subjected to identical stress cycling. The reason for this is that the onset and growth of fatigue damage depend upon the uniformity within the material and of the surface, and these are variable.

Rainflow counting is a method for extracting the important fatigue stress cycles from a complex signal. These are defined in terms of their range and mean (or maximum/minimum values), so that they can be used in fatigue life calculations. This is necessary as most materials fatigue data is presented in terms of constant stress amplitudes. Since the largest stress cycles are dominant in fatigue terms, the method extracts the largest overall cycle, follwed by ever decreasing ones. This can lead to cycles, and it is common to define a gate, or filter, to ignoe the low-level stress cycles that contribute only marginally to damage.

Figure 2.4 illustrates the method. The axes are stress (or strain), and the cycle is defined in terms of its stress (or strain) maximum and minimum. The top half shows the stress (or strain) history in time, and the bottom half shows how these cycles appear on the stress-strain chart. Cycles A–D, D–E, E–F, and F–A are significant in damage terms, whereas B–C and G–H are not. Figure 2.5 shows the typical appearance of a fatigue-induced failure. The 'rings' show the progressive growth of the fatigue crack, and the granular area is the final fracture surface.

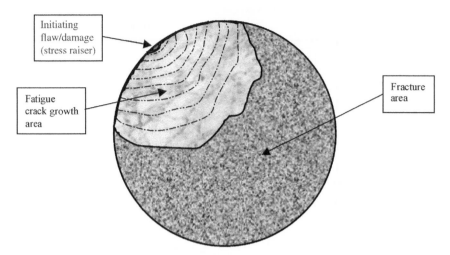

Figure 2.5   Typical fatigue failure (schematic)

Generally for metals the fatigue life, expressed as cycles to failure, is not affected by the rate at which stress cycling is applied. This is due to the fact that, since they are good thermal conductors, any energy converted to heat is readily conducted away, so there is little or no temperature rise. However, plastics generally are more likely to be locally heated by high rates of stress reversals, and this, coupled with their lower melting points and other properties, such as the glass transition temperature, can result in reduced fatigue lives at high cycle rates.

Composite materials, such as fibre-reinforced structural components, can be designed and manufactured to have tailored mechanical properties, since the stresses are transmitted primarily through the fibres rather than through the bulk material. Failure of composite components can be due to delamination or separation of the fibres, or fracture of the whole component.

2.2.2 Creep

Creep is the gradual increase in length of a component that is subjected to combined continuous or cyclic tensile stress and high temperature. Creep is a plastic deformation, which occurs when the material temperature exceeds about 50 percent of the melting point, on the absolute temperature scale. The effect is significant with components like turbine discs and blades in gas turbine engines, due to the combined very high temperatures and centrifugal forces. It has recently become a problem in electronics assemblies using surface mount components. Since solder melts at about 183°C, system operating temperatures are generally within the creep temperature range. Therefore, permanent deformation takes place due to the shear stresses imposed by thermal cycling. The deformation in turn can result in higher shear stresses, thus accelerating the fatigue mechanism.

References 2–7 provide excellent introductions to the material properties described above.

2.2.3 Vibration and shock

Components and assemblies can be subjected to vibration and shock inputs, during use, transport or maintenance. Vibration and shock can cause:

- Fracture due to fatigue, or due to mechanical overstress
- Wear of components such as bearings, connectors, etc.
- Loosening of fasteners, such as screws, bolts, etc.
- Leaks in hydraulic and pneumatic systems, due to wear of seals or loosening of connectors
- Acoustic noise (10–10 000 Hz).

Common vibration inputs are:

- Reciprocating or rotating machinery. The dominant vibration frequency (Hz) generated by rotating masses will be rpm/60
- Wheel vibration, on road and rail vehicles
- Aerodynamic effects on aircraft and missile structures
- Pressure fluctuations in hydraulic and pneumatic systems
- Acoustic noise

Vibration of a structure may occur at a fixed frequency, at different frequencies over time, or simultaneously over a range of frequencies. Vibration over a wide range of simultaneous frequencies is called *broad band* vibration. Vibration can occur in or about different linear and rotating axes.

The important units in relation to sinusoidal vibration are:

- Frequency (Hz)
- Displacement (mm), defined as peak or peak-to-peak values
- Velocity (m/s), defined as peak values
- Acceleration (m/s^2 or g_n), defined as peak values.

Every structure has one or more resonant frequencies, and if the vibration input occurs at these, or at harmonics, vibration displacements will be maximised. The locations at which zero vibration displacements occur are called *nodes*, and maximum displacement amplitudes occur at the *anti-nodes*. There may be more than one resonant frequency within the expected environmental range, and these may exist along different axes. There may also be more complex resonance modes, such as torsional or combinations of mechanical, acoustic, rotational or electromechanical modes. Sometimes simultaneously occurring resonances might be important, for example two components that vibrate in different modes or at different frequencies and in so doing impact one another. Common examples are electronic circuit boards, hydraulic pipes and vehicle panels.

The resonant frequency is proportional to the stiffness of the structure and inversely proportional to the inertia. Therefore, to ensure that resonant frequencies are well above any input vibrations that might be applied, structures must be sufficiently stiff, especially where there are relatively heavy parts, such as large components on circuit boards.

The vibration amplitude at any frequency is reduced by *damping*. Damping can also change the resonant frequency. Damping is provided in hydraulic and pneumatic systems by accumulators, in suspension and steering systems by mechanical dampers, and by using anti-vibration mountings for motors, electronic boxes, etc.

The pattern of vibration as a function of other parameters, such as engine speed, can be shown on a *waterfall plot*. Figure 2.6 is an example. Waterfall plots help to indicate the sources of vibration and noise. For example, a resonance that is independent of speed shows as a vertical line, and one that is generated at a particular speed shows peaks running horizontally. The peak heights (or colours on colour map displays) indicate the amplitude.

Since the topics of mechanical vibration and acoustic noise are closely related, they are sometimes considered and analysed together, along with the related concept of *harshness*, which is a subjective aspect, as *noise, vibration and harshness* (NVH). This approach is used particularly in the automotive industry.

Shock loads can cause vibration, though the amplitude is usually attenuated due to inherent or applied damping. Shock loads are only a particular type of vibration input: relatively high intensity and frequency, for short intervals.

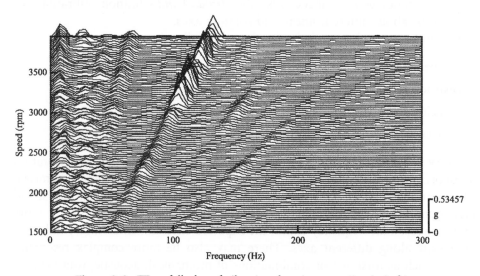

Figure 2.6   Waterfall plot of vibration data (courtesy Prosig Ltd.)

Reference 8 provides a comprehensive treatment of the subject. Testing methods for vibration and shock will be covered in Chapter 7.

2.3 TEMPERATURE EFFECTS

Failures can be caused by materials being subjected to high or to low temperatures. The main high-temperature failure modes are:

- Softening and weakening (metals, some plastics)
- Melting (metals, some plastics)
- Charring (plastics, organic materials)
- Other chemical changes
- Reduced viscosity or loss of lubricants
- Interaction effects, such as temperature-accelerated corrosion.

Low-temperature effects can include embrittlement of plastics, increasing viscosity of lubricants, condensation, and freezing of condensation or coolants.

Most temperature effects are deterministic (melting points, condensation temperatures, freezing points, viscosities). Effects such as these are not cumulative, so time and numbers of temperature cycles do not directly affect reliability. However, secondary effects might be cumulative, for example the effects of lubricant viscosity on rate of wear.

All materials have a *thermal coefficient of expansion* (TCE). If two components with different TCEs are attached to one another, or two attached components can experience different temperatures, then mechanical stresses will be set up. An important example of this situation is the attachment of electronic components to circuit boards or other substrates, particularly surface-mounted integrated circuit packages that are attached to the circuit board by solder joints around the package periphery, as described in Chapter 8. When the IC is powered and operated heat is generated, so the package temperature rises. The heat is transferred through the package to the circuit board (which might include a 'heat plane', to improve heat dissipation). The thermal resistance of the heat flow path from the package to the eventual heat sink will result in the package being hotter than the board, and in the board temperature rise lagging that of the package. If power is cycled, the temperature differences will be also. This will result in cyclic shear stresses being imparted to the solder joints. The magnitude of these stresses can lead to fatigue failures, in the form of cracks running through the joint. These in turn can cause electrical failure, usually of an intermittent nature, after a sufficient number of cycles. This type of failure is particularly important in electronic systems that must withstand many on–off cycles, such as engine control systems. If the systems are also subjected to vibration, the combined effects of thermal and vibration cycling can be highly interactive.

Chemical reactions, gaseous and liquid diffusion and some other physical processes are accelerated by increasing temperature. *Arrhenius' Law* expresses this phenomenon:

$$R = K \exp(-E/kT)$$

where:

R = process rate
K = constant
E = activation energy for the process
k = Boltzmann's constant
T = temperature (K).

Typically, chemical process rates increase by a factor of 2 for every 10–20°C rise in temperature. An important group of processes that can be thermally accelerated is corrosion, particularly rusting of iron and steel. We will discuss corrosion later.

2.4   WEAR

Wear is the removal of material from the surfaces of components as a result of their movement relative to other components or materials. Wear can occur by a variety of mechanisms, and more than one mechanism may operate in any particular situation. The science and methods related to understanding and controlling wear in engineering comprise the discipline of *tribology*. The main wear mechanisms are described below.

• *Adhesive wear* occurs when smooth surfaces rub against each other. The contact load causes interactions between the high spots on the surfaces, and the relative motion creates local heating and dragging between the surfaces. This results in particles being broken or scraped off the surfaces, and loose particles of wear debris are generated.

• *Fretting* is similar to adhesive wear, but it occurs between surfaces subject to small oscillatory movements. The small movements prevent the wear debris from escaping from the wear region, so the particles are broken up to smaller sizes and might become oxidised. The repeated movements over the same parts of the surface also result in some surface fatigue, and corrosion also contributes to the mechanism.

• *Abrasive wear* occurs when a relatively soft surface is scored by a relatively hard surface. The wear mechanism is basically a cutting action, often with displacement of the soft material at the sides of grooves scored in the soft material.

• *Fluid erosion* is caused to surfaces in contact with fluids, if the fluid impacts against the surfaces with sufficient energy. For example, high-velocity fluid jets can cause this type of damage. If the fluid contains solid particles the wear is accelerated.

- *Cavitation* is the formation and violent collapse of vacuum bubbles in flowing liquids subject to rapid pressure changes. The violent collapse of the vacuum bubbles on to the material surfaces causes fluid erosion. Pumps, propellers and hydraulic components can suffer this type of damage.
- *Corrosive wear* involves the removal of material from a surface by electrolytic action. It is important as a wear mechanism because other wear processes might remove protective films from surfaces and leave them in a chemically active condition. Corrosion can therefore be a powerful additive mechanism to other wear mechanisms.

References 9 and 10 are excellent introductions to tribology and wear.

2.5   CORROSION

Corrosion affects ferrous and some other non-ferrous engineering metals, such as aluminium and magnesium. It is a particularly severe reliability problem with ferrous products, especially in damp environments. Corrosion can be accelerated by chemical contamination, for example by salt in coastal or marine environments.

The primary corrosion mechanism is oxidation. Some metals, particularly aluminium, have oxides which form as very hard surface layers, thus providing protection for the underlying material. However, ferrous alloys do not have this property, so oxidation damage is cumulative.

Galvanic corrosion can also be a problem in some applications. This occurs when electromotive potentials are built up as a result of dissimilar metals being in contact and conditions exist for an electric current to flow. This can lead to the formation of intermetallic compounds and the acceleration of other chemical action. Also, *electrolytic* corrosion can occur, with similar results, in electrical and electronic systems when induced currents flow across dissimilar metal boundaries in the presence of an electrolyte, such as water with dissolved salts. This can occur, for example, when earthing or electrical bonding and material protection are inadequate. Electrolytic corrosion attacks the most electrically active element in the circuit.

Stress corrosion is caused by a combination of tensile stress and corrosion damage. Weakening occurs at crack tips, where the metal is in a chemically active state and where the high temperatures generated accelerate further chemical action. Thus the combined effect can be much faster than either occurring alone.

Reference 11 describes corrosion in detail.

2.6   HUMIDITY AND CONDENSATION

Damp environments can cause or accelerate failure processes such as corrosion and mould growth. Temperature and humidity are closely related, humidity being inversely proportional to temperature until the dewpoint is reached, below which moisture condenses onto surfaces. Liquid water can cause further failures, including:

- Chemical corrosion, if contamination is also present
- Electrolytic corrosion, by providing an electrolyte
- Short-circuiting of electrical systems, particularly within connectors.

Plastics are generally hygroscopic, that is, they absorb moisture, whether above or below the dewpoint. Therefore any components that are encapsulated in plastics, particularly electronic components and assemblies, are in principle prone to moisture ingress. This presented until fairly recently a major limitation on the application of plastic-encapsulated components, since they suffered corrosion of the aluminium conductor metallisation when used in high-humidity environments. Their use in military and aerospace systems was banned. However, modern components such as integrated circuits have much improved protection against moisture, due to better control of the chip's surface protective layer and control of the plastic material purity and the encapsulating process, so that today there are few limitations on their application, and moisture-related failures are very rare.

2.7 MATERIALS AND COMPONENTS SELECTION

Selection of appropriate materials is an important aspect of design, and it is essential that designers are aware of the relevant properties in the application environments. With the very large and increasing range of materials available this knowledge is not easy to retain, and designers should obtain data and application advice from suppliers as well as design handbooks and other databases. A few examples of points to consider in selecting and testing engineering materials are given below. The list is by no means exhaustive, and it excludes obvious considerations such as strength, hardness, flexibility, etc., as appropriate to the application.

2.7.1 Metals

- Fatigue resistance
- Corrosion environment, compatibility
- Surface protection methods
- Electrochemical (electrolytic, galvanic) corrosion if dissimilar metals in contact.

2.7.2 Plastics, rubbers

- Resistance to chemical attack from materials in contact or in the local atmosphere (lubricants, pollutants, etc.)
- Temperature stability (dimensional, physical), and strength variation at high and low temperature

- Sensitivity to ultraviolet radiation (sunlight)
- Moisture absorption (all plastics are hygroscopic).

2.7.3   Fasteners

A huge range of different fastening methods and systems is available, including rivets, bolts and nuts, clamps, adhesives, etc. Fasteners can loosen under vibration or as a result of temperature cycling. Fasteners can fail due to fatigue, and fatigue cracks can start at holes for rivets and bolts.

Bolts and nuts can be combined with locking devices to prevent loosening. These include deformable plastic inserts, spring washers, crush washers, split pin retainers, adhesives, locking wires, etc. The integrity of many locking devices can be degraded if they are used more than once.

Bolts and nuts used in some applications must be accurately torque-loaded to ensure that the correct holding force is applied, and that the fasteners are not over-stressed on assembly.

2.7.4   Adhesives

Adhesives are used for many assembly operations, including aircraft and other vehicle structures, electronic component mountings onto heat sinks, locking of bolts and nuts, etc. The most commonly used industrial adhesives are epoxy plastics and cyanoacrylics. Epoxies are two-component adhesives which must be mixed shortly before use. Cyanoacrylics are contact adhesives that form an instant bond. Other adhesive compounds and systems include elastomerics (used in applications such as vibration isolation) and adhesive tapes.

All adhesives require careful preparation and cleaning of the surfaces to be bonded, and they all have limitations in relation to the kinds of materials they can bond. Adhesives also have temperature limits, and generally cannot withstand temperatures above 200°C.

Testing of systems that are assembled with adhesives should include the effects of variation in application: mixing (for epoxies), preparation, cleaning, vibration, temperature extremes and cycling, etc.

2.7.5   Welding and soldering

Metals can be joined by welding, and several different welding methods are used, depending on the materials and the application. Steel structures are welded with electric arcs or oxy-acetylene gas torches. Alloys such as those of aluminium and magnesium, which burn in oxygen, are arc-welded in inert gas (argon). Car assemblies are spot-welded by robots applying pressure and high electric current to form resistance welds. Surfaces can also be welded by friction (high pressure and vibration, including ultrasonic welding of gold wire bonds on microelectronic assemblies).

Tin–lead solder is by far the most common method for connecting electronic and electrical components within systems. It also serves as a structural connection. Soldering for electronics assembly is described in Chapter 3.

Testing of welds and solders should include effects of variation, as well as cyclic stress applications that could lead to fatigue (vibration, thermal, etc.).

2.7.6 Seals

Seals are used to prevent leaks in systems such as water, oil hydraulic and pneumatic components and pipe connections, around rotating shafts and reciprocating actuator rams, and to protect items in sealed containers. Special seals include those to block electromagnetic radiation from or into electronic equipment enclosures.

The effectiveness of seals is always influenced by control of assembly operations, and often also by maintenance. They are always affected by usage (wear, erosion, etc.), so they tend to degrade over time and use.

Testing of seals and systems that include them should include the effects of variation in assembly and maintenance, and the effects of deterioration due to wear, etc.

Reference 12 is a good introduction to engineering seals.

Reference 3 is an excellent source of information on material selection.

References

1. O'Connor, P. D. T., 1995, *Practical Reliability Engineering* (3rd edn), John Wiley & Sons, Ltd.
2. Gordon, J. E., 1968, *The New Science of Strong Materials*, Penguin.
3. Crane, F. A. A. and Charles, J. A., 1989, *Selection and Use of Engineering Materials* (2nd edn), Butterworths.
4. Carter, A. D. S., 1997, *Mechanical Reliability and Design*, Macmillan.
5. Hertzberg, R. W., 1976, *Deformation and Fracture Mechanics of Engineering Materials*, John Wiley & Sons, Ltd.
6. Dowling, N. E., 1993, *Mechanical Behaviour of Materials* (2nd edn), Prentice-Hall.
7. Bannantine, J. A., Comer, J.J. and Handrock, J. L., 1989, *Fundamentals of Metal Fatigue Analysis*, Prentice-Hall.
8. Harris, C. E. and Crede, C. E., 1976, *Shock and Vibration Handbook*, McGraw-Hill.
9. Summers-Smith, J. D., 1994, *An Introductory Guide to Industrial Tribology*, Mechanical Engineering Publications.
10. Lipson, C., 1967, *Wear Considerations in Design*, Prentice-Hall.
11. Uhlig, H. H. and Revie, R. W., 1985, *Corrosion and Corrosion Control* (3rd edn), John Wiley & Sons, Ltd.
12. Summers-Smith, J. D. (ed.), 1992, *Mechanical Seal Practice for Improved Performance* (2nd edn), Mechanical Engineering Publications.

3

Electrical and Electronics Stress, Strength and Failure

3.1 INTRODUCTION

Electronic components and circuits can be caused to fail by most of the same mechanisms (fatigue, creep, wear, corrosion, etc.) described in the previous chapter. Fatigue is a common cause of failure of solder joints on surface-mounted components and on connections to relatively heavy or unsupported components such as transformers, switches and vertically mounted capacitors. Wear affects connectors. Corrosion can attack aluminium conductors on integrated circuits, connectors and other components. Electrical and thermal stresses can also cause failures that are unique to electronics. The main electrical stresses that can cause failures of electrical and electronic components and systems are current, voltage and power. For all of these failure modes there are strong interactions between the electrical and thermal stresses, since current flow generates heat.

As stated in the previous chapter, we need to understand and apply the physical laws and other factors that influence stresses and strengths in order to prevent failures. Our understanding of these derives from basic knowledge and experience, but in practical electrical and electronic engineering this is often even more incomplete and uncertain than the mechanisms discussed earlier. Therefore testing is necessary. Determining the most appropriate tests and interpreting the results depends upon sound appreciation of the science involved. In this chapter we will review the causes of failures of electrical and electronic components and systems, in sufficient detail to provide a basis for effective testing.

It is important to appreciate the fact that the great majority of electronic component types do not have any mechanisms that will cause degradation or failure during storage or use, provided that they are:

- properly selected and applied, in terms of performance, stress and protection;

- not defective or damaged when assembled into the circuit;

- not overstressed or damaged in use.

The quality of manufacture of modern electronic components is so high that the proportion that might be defective in any purchased quantity is typically of the order of less than 10 per million for complex components like ICs, and even lower for simpler components. Therefore the potential reliability of well-designed, well-manufactured electronic systems is extremely high, and there are no practical limits to the reliability that can be achieved with reasonable care and expenditure. This, however, has not always been the case. In the 1950s, when military electronic systems were becoming more complex and semiconductor components were in their infancy, component failures were the major causes of systems failures. As a result, a method for predicting the reliability of electronic systems was developed by the US Department of Defense. This became Military Handbook 217. The foundations of MIL-HDBK-217 are that:

- All system failures are caused by component failures, and all component failures cause system failures.

- All component types have a property of 'failure rate' (expressed as failures per million hours (FPMH) (λ) of system operation.

- The component failure rates are proportional to applied thermal and electrical stresses, and related to other aspects such as application environments and sources of supply, in ways that can be derived from system failure data and statistical regression, then applied credibly to future applications.

None of these is correct. (The relationship between reliability and temperature will be discussed later.) MIL-HDBK-217 has for some time been a controversial method, and its use has been forbidden by NASA, the US Army and the UK Ministry of Defence. It is not used by most electronics companies outside the defence and aerospace industries. Nevertheless it is still used by some organisations and companies. Also, similar methods have been developed and are applied by some telecommunications businesses (Bellcore in the USA and British Telecom), and others.

Methods such as these should never be used in modern electronics design and development, including test planning or the interpretation of test results. They provide highly misleading (and pessimistic) predictions of potential reliability, without any compensating advantages. Reference 1 discusses the methods in more detail.

3.2 STRESS EFFECTS

3.2.1 Current

Electrical currents cause the temperatures of conductors to rise. If the temperature reaches the melting point, the conductor will fuse. (Of course, fuses are used as protective devices to prevent other, more serious failures from occurring.)

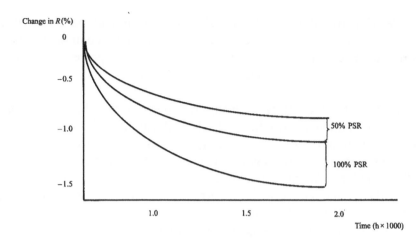

Figure 3.1 Parameter drift characteristics: carbon resistor + 70°C (PSR: power stress ratio = applied power/rated power)

Heat in conductors is transferred to other components and to insulation materials, primarily by conduction and convection, so thermal damage can be caused to these.

High currents can also cause component parameter values, such as resistance, to drift over time. This effect is also accelerated by high operating temperatures. Figure 3.1 shows examples of this.

Electric currents also create magnetic fields. If oscillating, they can generate acoustic noise and electromechanical vibration. We will discuss electrical interference effects later.

3.2.2 Voltage

Voltage stress is resisted by the dielectric strength of the material between the different potentials. The most common examples are the dielectric material between capacitor plates, and the insulation (air or other insulator) between conductors. Potential differences generate currents in conductors and components, and if the current-carrying capacity is insufficient the conductor or component will fail, in which case the failure mechanism is current, though the cause might be that the voltage is too high. For example, an integrated circuit might fail due to current overstress if a high electrostatic voltage is accidentally applied to it, and a 110 V appliance might fail for the same reason if connected to a 240 V supply.

High voltage levels can be induced by:

- *Electrostatic discharge* (ESD), caused by charge accumulation on clothing, tools, etc.

- Other electrical overstress, such as high voltage transients on power lines, unregulated power supplies, circuit faults that lead to components being overstressed, accidental connection of high voltages to low power components, etc. This is referred to as *electrical overstress* (EOS).

Another effect of voltage stress is *arcing*, which can occur whenever contacts are opened, for example in switches and relays. Arcing can also occur between brushes and commutators of motors and generators. Arcing generates electromagnetic noise, and also progressively damages the contact surfaces, leading eventually to failure. Arcing can also cause damage to components such as bearings, if the component provides a current path due to incorrect design or maintenance of the electrical circuit.

Arcing can be reduced by using voltage suppression components, such as capacitors across relay or switch contacts. Arcing becomes more likely, and is more difficult to suppress, if atmospheric pressure is reduced, since the dielectric constant of air is proportional to the pressure. This is why aircraft and spacecraft electrical systems operate at relatively low voltage levels, such as 28 V DC and 115 V AC.

Corona discharge can occur at sharp points at moderate to high voltage levels. This can lead to dust or other particles collecting in the area, due to ionisation.

Some components can fail due to very low or zero current or voltage application. Low-power relay contacts which pass very low DC currents for long periods can stick in the closed position due to cold welding of the contact surfaces. Electrical contacts such as integrated circuit socket connectors can become open due to build-up of a thin dielectric layer caused by oxidation or contamination, which the low-voltage stress is unable to break down.

3.2.3 Temperature

The Arrhenius model (Chapter 2) has been used to describe the relationship between temperature and time to failure for electronic components, and is the basis of methods for predicting the reliability of electronic systems, such as US MIL-HDBK-217. However, this is an erroneous application, since, for the great majority of modern electronic components, most failure mechanisms are not activated or accelerated by temperature increase. We will discuss other stress and failure aspects of electrical and electronics systems in more detail later. For the great majority of electronic components the true relationship between temperature and failure is as shown in Figure 3.2. Most electronic components can be applied at temperatures well in excess of the figures stated in databooks. For example, databook package temperature limits for industrial-grade plastic-encapsulated integrated circuits and transistors are typically 85°C, and for ceramic or metal packaged devices 125°C. These do not, however, relate to any physical limitations, but are based more on the conventions of the industry.

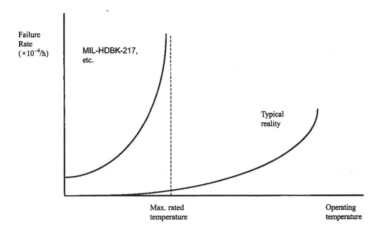

Figure 3.2 Failure rate vs. temperature for electronic components

High temperatures can cause damage or incorrect operation in electrical and electronic components. Wire coils in components such as inductors, solenoids, transformers and motors can become short-circuited if high temperature, caused by high current or high ambient temperature, causes insulation charring.

Low temperatures can also cause components to fail, usually due to parametric changes in electrical characteristics. Typical low-temperature limits for most components are −20°C to −60°C. However, such failures are usually reversible, and correct function is regained if the temperature rises.

Repeated temperature changes can be more damaging than continuous operation at high temperatures. Components such as power transistors and processors can suffer fatigue damage of internal wire bonds and of attachments to heat sinks. Temperature changes also cause fatigue damage and creep deformation of solder joints on surface-mounted electronic components, as described in Chapter 2.

References 1 and 2 discuss temperature effects on electronics in detail.

3.2.4 Power

Electrical power generates heat ($W = I^2 R$). 'Active' electronic components, such as transistors, integrated circuits and amplifiers, all generate heat, and therefore increased temperature. So do components like coils, voltage-dropping resistors, etc. The steady-state component temperature will be the sum of the ambient temperature and the temperature generated by internal heating. The internal heat is dissipated by conduction through the device connections and mountings, and the thermal resistance between the active part of the device and the ultimate heat sink will determine the steady-state temperature. We discussed the effects of temperature on electronic device reliability earlier.

Some passive devices, such as resistors and capacitors, are also susceptible to failure due to power stress, if it causes overheating (rather than fusing due to excess current). Power stress over long periods can also cause drift of parameter values such as resistance or capacitance. Application instructions for such components generally include power stress limits.

Power stress cycling can lead to failure due to induced thermal cycling and thus fatigue, as described earlier.

3.3 COMPONENT TYPES AND FAILURE MECHANISMS

The main categories of electronic component types and their most common failure modes are described in the sections below.

3.3.1 Integrated circuits (ICs)

Classes of ICs

ICs are not really 'components'. They are in fact subsystems, containing tens to millions of transistors, capacitors and other discrete components within the IC structure. Conventionally, ICs are classified as follows: Small Scale Integration (SSI) up to 100 logic gates; Medium Scale Integration (MSI) up to 1000 gates; Large Scale Integration (LSI) up to 10 000 gates; Very Large Scale Integration (VLSI) up to 100 000 gates; Ultra Large Scale Integration (ULSI) more than 100 000 gates. Currently (2000) microprocessors contain over 10 million transistors. VLSI/ULSI chips possess very many or even an unlimited range of potential functions.

When integrated circuits were first produced in the 1970s and 1980s they were mainly fairly simple analogue and digital circuits with defined functions (op-amps, adders, flip-flops, logic gates, etc.), and they were produced to closely defined generic specifications. For example, a 7408 is a two-input AND gate, and it might have been manufactured by several suppliers. In parallel with the rapid growth in complexity and functionality in the years since, several different classes of IC have been developed. The classes that are available today include:

- 'Standard' ICs. These are the components that appear in generic specifications or in manufacturers' catalogues. Examples are logic, memories, microprocessors, analogue to digital converters, signal processing devices, op-amps, etc.

- Programmable logic devices (PLDs), field programmable gate arrays (FPGAs). These are standard circuits which can be 'programmed' by selective opening of fusible links.

- Application-specific ICs (ASICs). These are ICs that are designed and made for particular applications, such as mobile telephones, games consoles, etc. They are not listed in catalogues. Many modern systems utilise a mixture of standard ICs and ASICs.

- Mixed signal ICs. These have both digital and analogue circuits integrated into the same chip.

- Microwave monolithic ICs (MMICs). These are used in systems such as telecommunications, and contain microwave radio circuits.

- Complex multi-function devices ('*system on a chip*'), which might contain mixed technologies, such as processing, memory, analogue–digital conversion, optical conversion, etc., and a mixture of new and 'legacy' circuit designs.

Packaging for ICs

There is also a wide variety of packaging methods for the IC chips. The *dual in line package* (DIP) was the most common standard configuration until the early 1980s. DIP technology uses conductor pin spacing of 0.1 inch (2.5 mm). Nowadays there are many types of *surface-mount technology* (SMT) packages. SMT allows far more electronics to be fitted into any area or volume, since the packages cover an area about 0.2 times that of an equivalent DIP and are thinner. Other package configurations are *small outline ICs* (SOICs) and *quad flat-packs* (QFPs). SMT connector spacing is typically 1 mm, with up to 120 connections on large QFPs. *Pin grid array* (PGA) and *ball grid array* (BGA) packages for large integrated circuits use an array of connections on the underside of the package, at spacings of 1 mm or less. Modern array packages have up to 500 connections. For compact systems such as mobile phones, portable PCs, etc., *chip scale packaging* (CSP) is used. *Multi-chip modules* (MCMs) are packages that contain a number of ICs.

Early DIP devices were available in ceramic or metal (hermetic) packages, as well as plastic (epoxy) encapsulated devices (PEDs). Hermetic devices were specified for military, aerospace and other high-temperature or high-humidity applications, since PEDs were considered unsuitable for such environments. PEDs are, however, much less expensive, and the encapsulation materials and processes have been improved to such an extent that nearly all modern devices are PEDs.

The mechanical and thermal responses of the package and connection structure and materials, as discussed in Chapter 2, are major considerations in testing electronic assemblies and systems that use SMT devices.

IC failure modes

The main failure modes of ICs are as follows.

- *Electrical overstress/electrostatic damage* (EOS/ESD). ICs are susceptible to damage from high voltage levels, which can be caused by transient events such as switching or electrostatic discharge from people or equipment. Most integrated circuits contain built-in EOS/ESD protection circuits, which will typically protect them against short-duration overstress conditions (typically up to 1000 V and 500 µJ).

- *Latchup* is the creation of a low-resistance path between the power input and ground in an integrated circuit. CMOS ICs are prone to this failure mechanism when subjected to transient voltage overstress caused by ESD or other transient pulses from circuit operation, test equipment, etc. The effect is permanent complete failure of the device.

- *Electromigration* (EM) is a failure mechanism that is becoming increasingly important as the aluminium conductor strips (*metallisation*) on integrated circuits are made to extremely narrow dimensions (currently down to about 0.25 micron across and 0.1 micron deep). Such cross-sectional areas mean that the current density (amps/metre^2), even at the very low current and voltage levels within such circuits, can be very high. EM is the bulk movement of conductor material, at the level of individual metal crystals, due to momentum interchange with the current-carrying electrons. This can result in local narrowing of the conductor track, and thus increased local current density and eventual fusing. Also, the displaced material can form conducting whiskers, which can cause a short-circuit to a neighbouring track. The EM process can be quantified using *Black's Law*, which expresses the rate of electromigration as an exponential function of temperature of the conductor and the electric current. EM is an important failure mode in electronic systems which must operate for long periods, particularly if operating temperatures are high, such as in engine controls, spacecraft and telecommunications systems (repeaters, switches, etc.).

- *Time-dependent dielectric breakdown* (TDDB) is a failure mode of the capacitors within ICs caused by whiskers of conductive material growing through the dielectric (silicon dioxide), and eventually short-circuiting the device. The effect is accelerated by voltage stress and by temperature.

- *Slow trapping* is the retention of electrons in the interstitial boundaries between Si and SiO_2 layers in ICs. These cause incorrect switching levels in digital logic and memory applications. Susceptibility to slow trapping is primarily dependent on device manufacturing processes.

- *Hot carriers* are electrons (or holes) that have sufficient energy to overcome the energy barrier of the Si–Si and SiO_2 boundary, and become injected into the SiO_2. This phenomenon occurs in sub-micron ICs in which the electric field strengths can be sufficiently high. The effects are to increase switching times in digital devices and to degrade the characteristics of analogue devices. Hot carrier effects can be reduced by process design techniques and by circuit design, both to reduce voltage stress at critical locations.

- *Soft errors* are the incorrect switching of a memory cell caused by the passage of cosmic ray particles or alpha particles. Cosmic rays create such effects in circuits in terrestrial as well as space applications. Alpha particles are generated by trace heavy metal impurities in device packaging materials. The errors can be corrected by refreshing the memory.

- Processing problems in manufacture (diffusion, metallisation, wire bonding, packaging, testing, etc.) can cause a variety of other failure mechanisms. Most will result in performance degradation (timing, data retention, etc.) or complete failure.

References 3–6 describe microelectronic component reliability physics.

3.3.2 Discrete semiconductors

- Processing problems in manufacture (diffusion, surface condition, metallisation, wire bonding, packaging, testing, etc.) can cause a variety of failure mechanisms. Most will result in performance degradation or complete failure.

- For power devices the uniformity and integrity of the thermal bond between the chip and the package is an important feature to ensure reliability at high power and thermal stress.

3.3.3 'Passive' components

- Resistors, capacitors, inductors and other components can fail due to fabrication problems, damage on assembly into circuits, ESD, and other causes. These usually cause the component to become open-circuit or high resistance.

- Component parameter values can be out of tolerance initially, or parameter values can drift over time due to applied stresses, as described earlier.

- Components can be electrically 'noisy', due to intermittent internal contacts or impurities in materials.

Capacitors

- High-voltage overstress generally causes the capacitor to become open-circuit. High-voltage/high-power capacitors might even explode if they are short-circuited.

- Low-voltage, low-power capacitors, such as those built into integrated circuits as memory storage devices, can suffer long-term failures due to the mechanism of dendritic metal whisker growth through the dielectric.

- Capacitors that use a liquid or paste dielectric (electrolytic capacitors) degrade over time if no voltage stress is applied, and then fail by short-circuit when used. They must be 're-formed' at intervals to prevent this. Capacitors kept in storage, or units such as power supplies which contain such components and which are stored or kept idle for long periods, must be appropriately maintained and checked.

- Electrolytic capacitors are damaged by reverse or alternating voltage, and so must be correctly connected and if necessary protected by diodes. Miniature tantalum capacitors are also degraded by ripple on the applied voltage, so they should not be used with unsmoothed voltage levels.

3.3.4 Electro-optical components

Many modern systems use optical fibres, connectors and electro-optical (EO) components for data transmission. Optical frequencies permit very high data rates. A major reliability advantage of EO systems is that they do not create EMI and they are immune to it. EO components are also used to provide over-voltage protection on data lines, by converting electrical signals to optical signals and back again to electrical. Failure modes of EO components are:

- Breakage of optical fibres

- Misalignment of optical fibres at connections within connectors and to components (the connecting ends must be accurately cut across the length and aligned with one another)

- Degradation of light output from light-emitting diodes (LEDs).

Reference 7 describes reliability aspects of most types of electronic components, and Reference 8 describes thermal effects on microelectronics packages.

3.3.5 Solder

Tin–lead solder (typically Sn63Pb37, melting point 183°C) is by far the most common material used for attaching electronic components to circuit boards. However, because of the toxicity of lead, efforts are being made to develop lead-free soldering, using alloys of tin with silver, copper and other metals. These all have higher melting points, up to 235°C, which in turn can affect component reliability. Solder joints can be made manually on relatively simple circuits that do not utilise components with fine pitches (less than 2.5 mm) between connections. However, for the vast majority of modern electronic systems solder connections are made automatically. The main techniques used are:

- *Through-hole mounting and manual or wave solder.* Components are mounted onto circuit boards, with their connections inserted through holes.

DIP ICs are mounted this way, as are many types of discrete components. The hole spacing is typically 2.5 mm. In wave soldering, the boards are passed over a standing wave in a bath of liquid solder, so that each connection on the underside is immersed briefly in the wave. These methods are now used mainly for components and circuits which do not need to utilise the most compact packaging technologies, such as power circuits.

- *Surface-mount and infra-red or vapour phase solder.* The circuit boards are printed with solder paste at the component connection positions, and the SMT components are mounted on the surface by automatic placement machines. BGA solder connections are made by solder balls (typically 1 mm in diameter) being accurately positioned on the solder paste on the board, then the BGA package is positioned on top. The 'loaded' boards are passed through 'reflow' ovens that melt the solder for just long enough to wet the surfaces, before the solder solidifies. The ovens are heated either with infra-red radiation, or, in the more common vapour phase or convection ovens, with gas heated to above the solder melting point. In the latter, the latent heat of condensation of the gas is transferred to the solder, and heating is very even.

- Laser soldering is also used to a limited extent, but further developments are likely.

The different types of component and solder methods are sometimes used in combination on circuit boards.

A reliable solder joint must provide good mechanical and electrical connection. This is created by the formation of intermetallic alloys at the interfaces between the solder and the surfaces being joined. The most common reasons for solder joint failure are:

- Inadequate solder wetting of the surfaces to be joined, due to surface contamination or oxidation. Components must be carefully stored and protected before placement. Components should not be stored for long periods before assembly, and fine-pitch components should be handled only by placement machines.

- Insufficient heat (time or temperature) applied. The solder might be melted sufficiently to bond to one or both surfaces, but not enough to form intermetallics. Such joints will conduct but will be mechanically weak.

- Fatigue due to thermally induced cyclic stress or vibration, as described in Chapter 2.

- Creep due to thermally induced cyclic stress, as described in Chapter 2.

All of these can lead to operating failures that show up on production test or in service. Failures can be permanent or intermittent. Modern electronic circuits can contain tens of thousands of solder connections, all of which must be correctly made. Control of the solder processes is a major factor in ensuring quality and reliability, and inspection and test of joint quality is an important feature of modern production test systems, as will be discussed in Chapter 10.

Figure 3.3 shows examples of solder defects and failures. References 6 and 7 describe soldering methods, and References 9–11 describe solder processes and problems.

3.3.6   Cables and connectors

Electrical power and signals must be conducted within and between circuits. Cables and connectors are not usually perceived as high-technology or high-risk components, but they can be major contributors to unreliability of many systems if they are not carefully selected and applied.

The most common cable systems are copper wires in individual or multi-conductor cables. Multi-conductor cables can be round or flat (ribbon). Cable failure mechanisms include damage during manufacture, use or maintenance, and fatigue due to vibration or movement. Failures occur mainly at terminations such as connections to terminals and connectors, but also at points where damage is applied, such as by repeated bending around hinges. The failure modes are mostly permanent or intermittent open-circuit, or short circuit to earth. Cable runs should be carefully supported to restrict movement and to provide protection. Testing should address aspects such as the possibility of damage during assembly and maintenance, chafing of insulation due to vibration, fatigue due to vibration or other movement, etc.

The main types of connectors are circular multipin, for connecting round cables, and flat connectors for connecting ribbon cables and circuit boards. Individual wires are often connected by soldering or by using screw-down or push-on terminals. Low-cost connectors may not be sufficiently robust for severe environments (vibration, moisture, frequent disconnection/reconnection, etc.) or long-life applications. Connectors for important or critical applications are designed to be rugged, to protect the connector pin surfaces from moisture and contamination, and the connecting surfaces are gold plated.

In many modern electrical and electronic systems connectors contribute a high proportion, and often the majority, of failures in service. The most common failure modes are permanent or intermittent open-circuit due to damage or build-up of insulation on the mating surfaces due to oxidation, contamination or corrosion. Therefore it is important that they are carefully selected for the application, are protected from vibration, abuse, corrosion and other stresses, and that their failure modes are taken into account in the test programme.

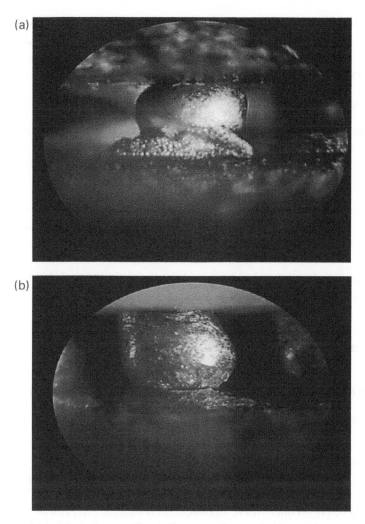

Figure 3.3 Solder failures: (a) solder ball, incomplete melt; (b) solder ball, poor bind, cracking (courtesy ERSA)

Data signals are also transmitted as pulses of light through optical fibre conductors and connectors. Light-emitting diodes operating in the infra-red part of the spectrum act as transmitting and receiving devices at the ends of the fibres. Special optical connectors are used to connect the ends of fibres, which must be accurately aligned and mated to ensure transmission. Optical fibres can break as a result of bending stresses.

3.3.7 Insulation

Insulation is as important in electrical and electronic systems as is conduction. All conductors, in cables and connectors and on circuit cards, must be insulated from one another. Windings in coils (solenoids, motors, generators, etc.) also require insulation. Insulation also provides protection against injury or death from human contact with high voltages.

Insulators can degrade and fail due to the following main causes:

- Mechanical damage, being trapped, chafed, cut, etc.

- Excessive temperature, causing charring and hence loss of dielectric strength. We are all familiar with the smell of burning shellac when an electrical appliance such as a drill or a microwave oven suffers a short-circuit in a motor or transformer coil.

- Embrittlement and then fracture, caused by exposure to high temperatures, UV radiation or chemical contamination. (Oil contamination can cause some cable insulator materials to swell and soften.)

- Rodent attack. Some insulation materials are bitten through by mice and rats which have access to agricultural machinery during the winter.

Degradation of insulation is nearly always a long-term phenomenon (10 or more years typically).

3.4   CIRCUIT AND SYSTEMS ASPECTS

3.4.1   Distortion and jitter

Distortion is any change in the shape of a waveform from the ideal. Distortion can be caused by several factors, including mismatched input and output impedances, crossover distortion in transistors, optical devices and op-amps, transistor saturation, interference (see below), thermal effects, etc. All waveforms (power, audio, HF to microwave signals, digital signals, etc.) can be affected, and the problems grow as frequencies increase. Circuit designs should minimise distortion, but it can cause failures, or it can be the symptom of component failures or parameter variations.

Jitter is a form of distortion that results in an intermittent variation of a waveform from its ideal position, such as timing, period or phase instabilities. It can affect the operation of high-speed circuits, particularly the timing of digital pulses, and can therefore cause corruption of signals and data.

3.4.2   Timing and interference

Timing is an important aspect of most digital electronic circuit design. To function correctly, the input and output voltage pulses to and from circuit elements must appear at the correct times in the logic sequences. This is relatively easy to arrange by design for simple circuits which operate at relatively low speed (clock rate or frequency). However, as speeds and complexity have increased, and continue to do so, it becomes increasingly difficult to ensure that every pulse occurs at the correct time and sequence. The integrity of the pulse waveform also becomes more difficult to assure at higher frequencies. Any digital circuit will have a speed above which it will begin to perform incorrectly. For example, in a production batch of microprocessors all are made to the same design and using the same processes, but they will have different maximum operating speeds, due to variations in the manufacturing processes, and the final functional test is used to determine which are sold as 266 MHz, 400 MHz, 800 MHz, etc., units. At higher assembly levels, such as telecommunications or control systems, further limitations on speed of operation can be caused by inductive and capacitive effects within the circuit and propagation delays along conductors.

Electromagnetic interference (EMI) is any disturbance of correct circuit operation due to the effects of changing electromagnetic fields. EMI is also called '*noise*'. EMI can be generated by sources external to the circuit, such as arcing or radio frequency emitters (radios, radars), lightning, etc. EMI can also be generated internally, by any conductor or component that is operating at high switching or AC frequencies. The operating frequencies of modern digital systems are in the VHF to UHF radio frequency range, and harmonics are also generated. The design objective is to ensure that all signals travel through the circuit conductors, but at such high frequencies there are inevitably radiated emissions, which can then be received by other conductors and components. Circuit design to prevent EMI is a difficult and challenging aspect of all modern system design. We will cover testing for EMI and its corollary, *electromagnetic compatibility* (EMC), in Chapter 8.

Reference 12 is a good introduction to EMI/EMC.

3.4.3 Intermittent failures

A large proportion of the failures of modern electronic systems are in fact of an intermittent nature. That is, the system performs incorrectly only under certain conditions, but not others. Such failures are most often caused by connectors that fail to connect at some times, such as under vibration or at certain temperatures, broken circuit card tracks that are intermittently open-circuit, tolerance build-up effects between component parameters, etc. It is not uncommon for more than 50 percent of reported failures of systems to be diagnosed on investigation as '*no fault found*' (NFF) or '*retest OK*' (RTOK), mainly due to the effects of intermittent failures. Worse, since the causes of the failures are mostly not detected, the faulty units are not repaired, and can cause the same system failure when reinstalled. These can therefore generate high costs of system downtime, repair work, provision of spare units, etc.

3.4.4   Other failure causes

There are many other causes of failure of electrical/electronic components and systems. It is impracticable to attempt to try to provide a comprehensive list, but examples include:

- Failure of vacuum devices (CRTs, light bulbs and tubes, etc.) due to seal failures

- Failures of light bulbs due to sputtering of filament material and eventual loss of current-carrying capability. This mechanism generates a quite specific life-limiting feature, typically about 1000 hours

- Failures due to non-operating environments, such as storage or standby conditions. Reference 13 covers these aspects.

References 14 and 15 provide introductions to electronics reliability.

References

1. Lall, P., Pecht, M. G. and Hakim, E. B., 1997, *Influence of Temperature on Microelectronics and System Reliability*, CRC Press.
2. McCluskey, P., Grzybowski, R. and Podlesak T. (eds), *High Temperature Electronics*, CRC Press.
3. Pecht, M. G., Radojcic, R. and Rao, G., 1999, *Guidebook for Managing Silicon Chip Reliability*, CRC Press.
4. Jensen, F., 1995, *Electronic Component Reliability*, John Wiley & Sons, Ltd.
5. Kraus, R., Hannemann, R. and Pecht, M. (eds), 1994, *Semiconductor Packaging: A Multidisciplinary Approach*, John Wiley & Sons, Ltd.
6. Lau, J. H., Wong, C. P., Prince, J. L. and Nakayama, W., 1998, *Electronic Packaging: Design, Materials, Process, and Reliability*, McGraw-Hill.
7. Bajenescu, T. J. and Bazu, M. I., 1999, *Reliability of Electronic Components*, Springer-Verlag.
8. Lau, J. H. (ed.), 1993, *Thermal Stress and Strain in Microelectronic Packaging*, Van Nostrand Reinhold.
9. Woodgate, R. W., 1996, *The Handbook of Machine Soldering: SMT and TH* (3rd edn), Wiley-Interscience.
10. Pecht, M. (ed.), 1993, *Soldering Processes and Equipment*, John Wiley & Sons, Ltd.
11. Brindley, K. and Judd, M., 1999, *Soldering in Electronics Assembly*, Newnes.
12. Ott, H. W., 1988, *Noise Reduction Techniques in Electronic Systems* (2nd edn), Wiley-Interscience.
13. Pecht, J. and Pecht, M. (eds), *Long-Term Non-Operating Reliability of Electronic Products*, CRC Press.
14. O'Connor, P. D. T., 1995, *Practical Reliability Engineering* (3rd edn), John Wiley & Sons, Ltd.
15. Pecht, M. (ed.), *Product Reliability, Maintainability and Supportability Handbook*, ARINC Research Corp.

4

Variation and Reliability

4.1 VARIATION IN ENGINEERING

Engineering products exist not only as tangible objects but also are described by abstract properties which represent their state and behaviour. Because the most useful products are made and used in large numbers (and have diverse properties) there is a need to reassure users about the values of these properties. The great majority of products are assembled, as the designer intends, from standard mass-produced components. Despite the care taken in designing and controlling the processes by which they are made, their actual values differ from the nominal values. Therefore the properties of the products also differ. They are also used in different conditions.

The differences, or *variations*, of parameters in engineering applications (machined dimensions, material strengths, transistor gains, resistor values, temperatures, etc.) are conventionally described in two ways. The first, and simplest, is to state maximum and minimum values, or *tolerances*. This provides no information on the nature, or shape, of the actual distribution of values. However, in most practical cases this is sufficient information for the creation of manufacturable, reliable designs.

The second approach is to describe the nature of the variation, using data derived from measurements. In this section we will briefly review the methods of statistics in relation to the practical aspects of describing and controlling variation in engineering.

4.1.1 Distributed variables

The first step in dealing with variable data is to plot the measured values against the numbers of measurements that fall within defined intervals, and so create a *histogram* (Figure 4.1).

The most important statistical properties of any variable quantity are:

1. The *central tendency*, which determines the nominal value of the variable, or the *location* of the data. The parameters that can be used to describe this are

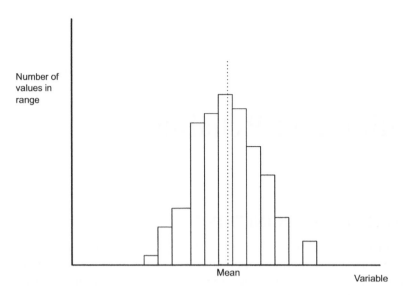

Number of
values in
range

Mean

Variable

Figure 4.1 Histogram

the *average* (the sum of all data values divided by the quantity of values), the *mean* (the value that lies on the centroid of the histogram), the *median* (the middle measured value), and the *mode* (the most likely value).

2. The spread of the variable quantity. This can be described by the *range* (the distance between the highest and lowest measured values), the *variance* (the sum of the squares of the differences between each measured value and the average, divided by one less than the number of measurements), and the *standard deviation* (SD) (the square root of the variance).

The data and the properties are called the *sample statistics*.

In the limit, assuming an infinite number of measurements and infinitesimal widths of the measurement intervals, the histogram will tend to a continuous curve. The equation describing such a continuous distribution is called the *probability density function*, or *pdf*, and the distribution properties are then called the *population parameters*. The mean of a pdf is usually denoted as μ (mu), and the SD as σ (sigma). The foundation of most analysis of variable data is to try to estimate the most likely population pdf and its parameters from the sample statistics available. Inevitably the more data available the more confident we can be in our estimates of the population's characteristics.

By far the most widely used 'model' of the nature of variation is the mathematical function known as the *normal* (or *Gaussian*) *distribution*. Figure 4.2 shows the shape of the normal distribution that best 'fits' the data of Figure 4.1. The reason why the normal distribution is commonly used as the model for

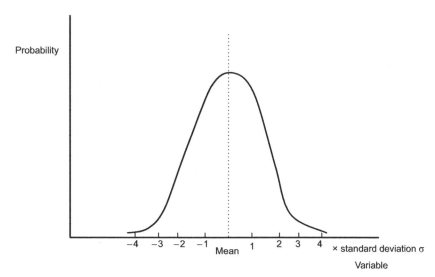

Figure 4.2   Normal (Gaussian) distribution

dealing with variation, in engineering as in many other fields, is that it can be shown that any variable quantity, whose total variation is the result of many separate contributing variables, will tend to the normal pdf, regardless of the nature of the contributing distributions. This is the *Central Limit Theorem*. It explains why natural processes such as human heights and rainfall amounts tend to be normally distributed. So, for example, if the variation of measured values of a machined dimension on a population of parts is the result of variation due to backlash in the machine bearings, vibration, tool wear, and measurement accuracy, then it is likely that the overall variation will tend to the normal distribution.

The normal distribution has another property that makes it convenient for statistical analysis. It is symmetrical about the mean, so that the probability of a measured value lying a certain distance either side is the same, and there are equal numbers of the population on either side. The mean coincides with the average and the median. The central limit theorem, and the convenient properties of the normal distribution, also explain why this particular function is taught as the basis of all statistics. It is common practice, in most applications, to assume that the variation being analysed is 'normal', then to determine the mean and SD of the normal distribution that best fits the data.

However, at this point we must stress an important limitation of assuming that the normal distribution describes the real variation of any process. The normal pdf has values between $+\infty$ and $-\infty$. Of course a machined component dimension cannot vary like this. The machine cannot add material to the component, so the dimension of the stock (which of course will vary, but not by much) will set an upper limit. The nature of the machining process, using

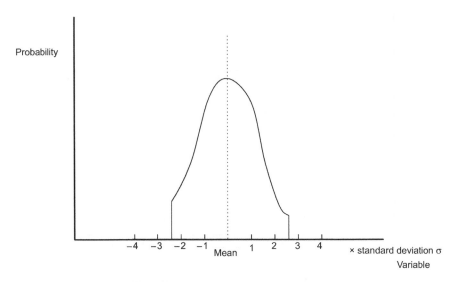

Figure 4.3 Curtailed normal distribution

gauges or other practical limiting features, will set a lower limit. Therefore the variation of the machined dimension would more realistically look something like Figure 4.3. Only the central part might be approximately 'normal', and the distribution will have been *curtailed*. In fact all variables, whether naturally occurring or resulting from engineering or other processes, are curtailed in some way, so the normal distribution, while being mathematically convenient, is actually misleading when used to make inferences well beyond the range of actual measurements, such as the probability of meeting an adult who is one foot tall.

There are other ways in which variation in engineering might not be 'normal'. These are as follows.

- There might be other kinds of selection process. For example, when electronic components such as resistors, microprocessors, etc. are manufactured, they are all tested at the end of the production process. They are then 'binned' according to the measured values. Typically, resistors that fall within ±2 percent of the nominal resistance value are classified as precision resistors, and are labelled, binned and sold as such. Those that fall outside these limits but within ±10 percent become non-precision resistors, and are sold at a lower price. Those that fall outside ±10 percent are scrapped. Because those sold as ±10 percent will not include any that are ±2 percent, the distribution of values is as shown in Figure 4.4. Similarly, microprocessors are sold as, say, 200 MHz, 400 MHz, 800 MHz, etc., operating speeds depending on the maximum speed at which they function correctly on test, having all been produced on the same process. The different maximum operating speeds are

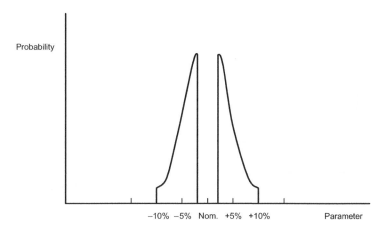

Figure 4.4 Effect of selection

the result of the variations inherent in the process of manufacturing millions of transistors and capacitors and their interconnections, on each chip on each wafer. The technology sets the upper limit for the design and the process, and the selection criteria set the lower limits. Of course, the process will also produce a proportion that will not meet other aspects of the specification, or that will not work at all.

- The variation might be unsymmetrical, or *skewed*, as shown in Figure 4.5. There are mathematical pdfs that represent such distributions, such as the lognormal and the Weibull distributions. However, it is still important to remember that these mathematical models will still represent only approxima-

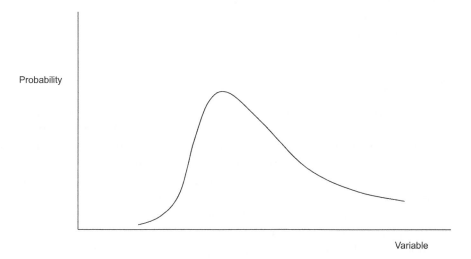

Figure 4.5 Skewed distribution

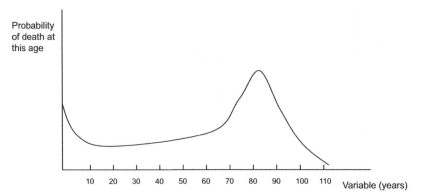

Figure 4.6 Bimodal distribution: typical human mortality

tions to the true variations, and the further into the tails that we apply them the greater will be the scope for uncertainty and error.

- The variation might be *multi-modal* (Figure 4.6), rather than unimodal as represented by distribution functions like the normal, lognormal and Weibull functions. For example, a process might be centred on one value, then an adjustment moves this nominal value. In a population of manufactured components this might result in a total variation that has two peaks, or a bimodal distribution. A component might be subjected to a pattern of stress cycles that vary over a range in typical applications, and a further stress under particular conditions, for example resonance, lightning strike, etc.

Variation of engineering parameters is, to a large extent, the result of human performance. Factors such as measurements, calibrations, accept/reject criteria, control of processes, etc. are subject to human capabilities, judgements and errors. People do not behave 'normally'.

4.1.2   Effects, causes and tails

Walter Shewhart, in 1931 (Reference 1), was the first to explain the nature of variation in manufacturing processes. Figure 4.7 illustrates four very different kinds of variation, which, however, all have the same means and SDs. These show clearly how misleading it can be to assume that any variation is 'normal' and then to make assertions about the population based upon the assumption.

In engineering (and in many other applications) we are really much more concerned with the effects of variation than with the properties and parameters. If, for example, the output of a process varied as in Figure 4.7(c), and the '$\pm n\sigma$' lines denoted the allowable tolerance, 100 percent would be in tolerance. If, however, the process behaved as in (a) or (d), a proportion would be outside

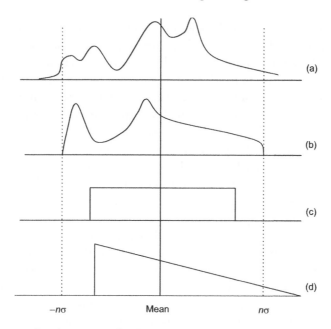

Figure 4.7 Four distributions with the same mean and SD (after W. A. Shewhart, Reference 1)

tolerance (only at the high end in the case of (d). Variation can have other effects. A smaller diameter on a shaft might lead to higher oil loss or reduced fatigue life. Higher temperature operation might make an electronic circuit shut down. A higher proportion of fast processors on a production batch would result in higher profit. We must therefore first identify the effects of variation (they are often starkly apparent), and determine whether and to what extent the effects can be reduced. This is not simply a matter of 'reducing SD'.

The effects of variation can be reduced in two ways:

1. We can compensate for the variation, by methods such as gauging or 'select on test' (this curtails the original variation), by providing temperature compensating devices, etc.

2. We can reduce the variation.

In both cases we must understand the cause, or causes, of the variation. Shewhart categorised manufacturing variation into *'assignable'* and *'non-assignable'* causes. (These are also referred to as *'special causes'* and *'common causes'*.) Assignable variation is any whose cause can be practically and economically identified and reduced. Non-assignable variation is that which remains when all of the assignable variation has been removed. A process in this state is *'in control*', and will have minimal variation. Note that these are practical criteria, with no mathematical basis. Shewhart developed the methods of *statistical process control* (SPC) around this thinking, with the emphasis on using the data and charting methods to identify and reduce assignable variation, and to keep processes in control. SPC charts are described in more detail in Chapter 12.

Again, statistical assumptions and methods are relatively unimportant and can be misleading if applied to this kind of situation. What must be emphasised instead is the motivation and practical knowledge necessary to identify and deal with the underlying causes. Later teachers, particularly W.E. Deming (Reference 2) and Genichi Taguchi (see Section 4.4.1) extended the ideas by demonstrating how reducing variation reduces costs and increases productivity, and by emphasising the management implications.

People such as insurance actuaries, clothes manufacturers and pure scientists are interested in averages and SDs: the behaviour of the bulk of the data. Since most of the sample data, in any situation, will represent this behaviour, they can make credible assertions about these population parameters. However, the further we try to extend the assertions into the tails, the less credible they become, particularly when the assertions are taken beyond any of the data. In engineering we are usually more concerned about the behaviour of variation at the extremes, than that near the average. We are concerned by high stresses, high and low temperatures, slow processors, weak components, etc. In other words, it is the tails of the distributions that concern us. We often have only small samples to measure or test. Therefore, using conventional mathematical statistics to attempt to understand the nature, causes and effects of variation in engineering can be misleading.

These are the aspects that matter in engineering, and they transcend the kind of basic statistical theory that is generally taught and applied.

Despite all of these reasons why conventional statistical methods can be misleading if used to describe and deal with variation in engineering, they are widely taught and used, and their limitations are hardly considered. Examples are:

- Most textbooks and teaching on SPC emphasise the use of the normal distribution as the basis for charting and decision-making. They emphasise the mathematical aspects, such as probabilities of producing parts outside arbitrary 2σ or 3σ limits, and pay little attention to the practical aspects discussed above.

- Typical design rules for mechanical components in critical stress application conditions, such as aircraft and civil engineering structural components, require that there must be a factor of safety (say 2) between the maximum expected stress and the lower 3σ value of the expected strength. This approach is really quite arbitrary, and oversimplifies the true nature of variations such as strength and loads, as described in the next section. Why, for example, select 3σ? If the strength of the component were truly normally distributed, about 0.1 percent of components would be weaker than the 3σ value. If few components are made and used, the probability of one failing would be very low. However, if many are made and used, the probability of a failure among the larger population would increase proportionately. If the component is used in a very critical application, such as an aircraft engine suspension bolt, this probability might be considered too high to be tolerable. Of course there are often other factors that must be considered, such as weight, cost, and the consequences of failure. The criteria applied to design of a domestic machine might sensibly be less conservative than for a commercial aircraft application.

• The so-called '*six sigma*' approach to achieving high quality is based on the idea that, if any process is controlled in such a way that only operations that exceed $\pm 6\sigma$ of the underlying distribution will be unacceptable, then only one per million operations will fail. The exact quantity is based on arbitrary and generally unrealistic assumptions about the distribution functions, as described above. ('Six sigma' entails other features, such as the use of a wide range of statistical and other methods to identify and reduce variations of all kinds, and the training and deployment of specialists called 'six sigma black belts'. It is not altogether a bad approach, but it is not the best, and it is heavily hyped by consultants. Reference 3 provides a practical description).

4.2 LOAD–STRENGTH INTERFERENCE

A common cause of failure results from the situation when the applied load exceeds the strength. Load and strength are considered in the widest sense. 'Load' might refer to a mechanical stress, a voltage or internally generated stresses such as temperature. 'Strength' might refer to any resisting physical property, such as hardness, strength, melting point, adhesion or voltage rating.
Examples are:

• A bearing fails when the internally generated loads (due perhaps to roughness, loss of lubrication, etc.) exceed the local strength, causing fracture, overheating or seizure.

• A transistor gate in an integrated circuit fails when the voltage applied causes a local current density, and hence temperature rise, above the melting point of the conductor or semiconductor material.

• A hydraulic valve fails when the seal cannot withstand the applied pressure without leaking excessively.

- A shaft fractures when the applied torque exceeds the strength.

Therefore, if we design so that strength exceeds load, we should not have failures. This is the normal approach to design, in which the designer considers the likely extreme values of load and strength, and ensures that an adequate safety factor is provided.

   Additional factors of safety may be applied, e.g. as defined in pressure vessel design codes or electronic component derating rules. This approach is usually effective. Nevertheless, some failures do occur which can be represented by the load–strength model. By our definition, either the load was then too high or the strength too low. Since load and strength were considered in the design, what went wrong?

   For most products neither load nor strength is fixed, but both are distributed statistically. This is shown in Figure 4.8(a). Each distribution has a mean value, denoted by $\bar{L}$ or $\bar{S}$, and a standard deviation, denoted by σ. If an item at the extreme weak end of the strength distribution is subjected to a load at the extreme high end of the load distribution, higher than the resisting strength, i.e. within the overlapping tails, failure will occur. This situation is shown in Figure 4.8(b).

   For distributed load and strength, we define two factors, the *safety margin* (SM):

$$SM = \frac{\bar{S} - \bar{L}}{\sqrt{(\sigma_S^2 + \sigma_L^2)}}$$

and the *loading roughness* (LR):

$$LR = \frac{\sigma_L}{\sqrt{(\sigma_S^2 + \sigma_L^2)}}$$

where:

$\bar{S}$ = mean strength
$\bar{L}$ = mean load
σ_S = SD of strength
σ_L = SD of load.

   The safety margin is the relative separation of the mean values of load and strength, and the loading roughness is the standard deviation of the load; both are relative to the combined standard deviation of the load and strength distributions. The safety margin and loading roughness allow us, in theory, to analyse the way in which load and strength distributions interfere, and so generate a probability of failure. By contrast, a traditional deterministic safety factor, based upon mean or maximum and minimum values, does not allow a probabilistic reliability estimate to be made. On the other hand, good data on

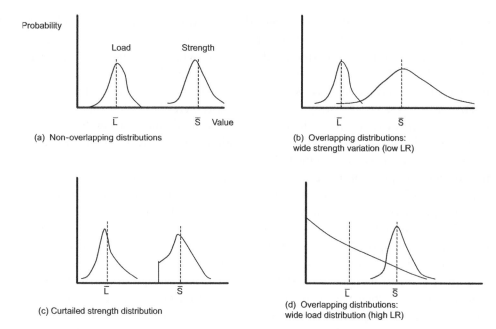

Figure 4.8   Distributed load and strength

load and strength properties, particularly at the extreme values represented by the overlaps of the tails, are very often not available. Other practical difficulties arise in applying the theory, and engineers must always be alert to the fact that people, materials and the environment will not necessarily be constrained to the statistical models being used.

Figure 4.8(a) shows a highly reliable situation: narrow distributions of load and strength, low loading roughness and a large safety margin. If we can control the spread of strength and load, and provide such a high safety margin, the design should be intrinsically failure-free. (Note that we are considering situations where the mean strength remains constant, i.e. there is no strength degradation with time. We will cover strength degradation later.) This is the concept applied in most designs, particularly of critical components such as civil engineering structures and pressure vessels. We apply a safety margin which experience shows to be adequate; we control quality, dimensions, etc., to limit the strength variations, and the load variation is either naturally or artificially constrained.

Figure 4.8(b) shows a situation where loading roughness is low, but due to a large standard deviation of the strength distribution the safety margin is low. Extreme load events will cause failure of weak items. However, only a small proportion of items will fail when subjected to extreme loads. This is typical of a situation where quality control methods cannot conveniently reduce the standard deviation of the strength distribution (e.g. in electronic device manufacture, where 100 percent visual and mechanical inspection is not always feasible). In this case deliberate overstress can be applied to cause weak items to fail, thus leaving a population with a strength distribution which is truncated to the left (Figure 4.8(c)). The overlap is thus eliminated and the reliability of the surviving population is increased. This is the justification for high stress burn-in of electronic devices, proof-testing of pressure vessels, etc. Note that the overstress test only destroys weak items. It must not cause weakening (strength degradation) of good items. A decreasing hazard or failure rate (DFR) is characteristic of this situation, since as weak items fail ('infant mortality'), the average strength of the surviving population increases and the failure rate decreases.

Figure 4.8(d) shows a low safety margin and high loading roughness due to a wide spread of the load distribution. This is a difficult situation from the reliability point of view, since an extreme stress event will cause a large proportion of the population to fail. Therefore, it is not economical to improve population reliability by screening out items likely to fail at these stresses. The options left are to increase the safety margin by increasing the mean strength, which might be expensive, or to devise means to curtail the load distribution. This is achieved in practice by devices such as current limiters in electronic circuits or pressure relief valves and dampers in pneumatic and hydraulic systems.

The examples illustrate some of the limitations of the statistical engineering approach to design. The main difficulty is that, in attempting to take account of variability, we are introducing assumptions that might not be tenable, e.g. by extrapolating the load and strength data to the very low probability tails of the assumed population distributions. We must therefore use engineering knowledge to support the analysis, and use the statistical approach to cater for engineering uncertainty, or when we have good statistical data. For example, in many mechanical engineering applications good data exists or can be obtained on load distributions, such as wind loads on structures, gust loads on aircraft or the loads on automotive suspension components. We will call such loading situations 'predictable'.

On the other hand, some loading situations are much more uncertain, particularly when they can vary markedly between applications. Electronic circuits subject to transient overload due to the use of faulty procedures or because of the failure of a protective system, or a motor bearing used in a hand power drill, represent cases in which the high extremes of the load distribution can be very uncertain. The distribution may be multimodal, with high loads showing peaks, for instance when there is resonance. We will call this loading situation 'unpredictable'.

It will not always be easy to make a definite classification; for example, we can make an unpredictable load distribution predictable if we can collect sufficient data. The methods described above are meaningful if applied in predictable loading situations. (Strength distributions are more often predictable, unless there is progressive strength reduction, which we will cover later.) However, if the loading is very unpredictable the failure probability estimates will be very uncertain. When loading is unpredictable we must revert to traditional methods. This does not mean that we cannot achieve high reliability in this way. However, evolving a reliable design is likely to be more expensive, since it is necessary either to deliberately overdesign or to improve the design in the light of experience. The traditional safety factors derived as a result of this experience ensure that a new design will be reliable, provided that the new application does not represent too far an extrapolation.

Instead of considering the distributions of load and strength, we can use discrete maximum/minimum values in appropriate cases. For example, we can use a simple lowest strength value if this can be assured by quality control. In other cases we might also assume that for practical purposes the load is curtailed, as in situations where the load is applied by a system with an upper limit of power, such as a hydraulic ram or a human operator. If the load and strength distributions are both curtailed, the traditional safety factor approach is adequate, provided that other constraints such as weight or cost do not make a higher risk design necessary.

The statistical engineering approach can lead to overdesign if it is applied without regard to real curtailment of the distributions. Conversely, traditional deterministic safety factor approaches can result in overdesign when weight or cost reduction must take priority. In many cases, other design requirements (such as for stiffness) provide intrinsic reliability. Probabilistic design techniques should therefore be used when it is necessary to assess the risk of failure in marginal or critical applications, and the test programme should be planned accordingly.

The theoretical basis and practical aspects of load–strength interference analysis are described in detail in References 4 and 5.

4.3   TIME-DEPENDENT VARIATION

Nearly all naturally occurring variation is, for practical purposes, constant in time. For example, human heights, IQs and death risks are variable (the first two pretty well follow the normal distribution, and death risks are bimodal and skewed, as shown in Figure 4.6), but these variations do not change between generations.

However, variations of parameters in engineering are not always constant in time. Manufacturing processes might vary from batch to batch, from day to day, from supplier to supplier, etc. Application environments might change if there is a change of use. A very important category of change is that of strength as a result of any or a combination of the many conditions that can cause weakening during use. In Chapter 2 we considered the major causes of progressive weakening: material fatigue caused by cyclic stress reversals, wear, corrosion, insulation breakdown, etc. In these cases the initial strength might be adequate, but it eventually deteriorates to the point that it can no longer withstand high values of applied stress. Figure 4.9 shows how the strength and load distributions might progress over time (or, in the case of fatigue, over the number of applied load cycles). The weakest components begin to fail after time t', when they are subjected to loads at the high end of the load distribution, and the rate at which failures occur increases with further application. This is typical of components subject to fatigue loading, but the same general picture applies to any mechanism that generates progressive weakening. Note that the distribution of applied load does not usually change, but the mean value of the strength distribution reduces, while the spread increases. This model is quite well understood for fairly simple fatigue-loading design situations, when the load and initial strength parameters and their distributions are known. Even so, there is still considerable uncertainty regarding when a component will eventually fail, typically over about an order of magnitude of stress cycles. We can say that the life, or time to failure, of relatively simple load–strength applications is 'predictable'. However, in more complex situations, for example when the load distribution is very uncertain, or when strength degradation is caused by other mechanisms or combinations such as fatigue and corrosion, the time to failure is 'unpredictable'. These terms do not, of course, represent definitive categories, but portions of a wide spectrum of uncertainty when dealing with time-dependent strength. The time-dependent aspects generate a further order of uncertainty relative to situations in which variables do not change over time.

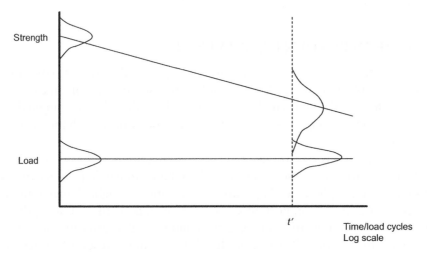

Figure 4.9 Load and strength variation over time

4.4 MULTIPLE VARIATIONS AND STATISTICAL EXPERIMENTS

So far we have considered situations in which only one parameter is variable, or two (load and strength). However, in many engineering situations there are several variable parameters. A system such as a pump or an electronic circuit will contain a number of components, each with variable dimensions or parameters, and these might also vary as a result of temperature or other stress. Some of the variables might have significant effects, others less. Some of the variations might be time-dependent. A particularly important aspect of multi-variable systems is that some of the variables might have effects that are *interactive*, so that their combined effects are significant. Sometimes interaction effects can be stronger than any single effect. Examples might be the combined effect of two modes of vibration, or of vibration and moisture, on the rate of fatigue damage, or the combined effects of two or more electrical parameter variations on the performance of an electronic circuit.

We can explore the effects of multiple variations by performing a series of traditional experiments, in which all variables are held constant except the one whose effect is being studied, and then repeating the procedure for the other variables. Those variables which have significant effects, and the nature of the effects, will eventually be determined. However, testing for the effect of one variable at a time can be very time-consuming and expensive. Also, such traditional one-at-a-time experiments *cannot detect interaction effects*.

R.A. Fisher invented the technique known as *statistical design of experiments* (DoE) to analyse the effects of variable inputs on the yields of crops. Here the variables are different fertilisers, trace elements, soil conditions, etc. In this kind of situation there are no scientific theories on which mathematical models of cause and effect can be based. The researchers know what the variables are, but they do not know how large their effects are, or whether the effects are interactive. They cannot therefore calculate the effects or simulate the system.

In a statistically designed experiment, a range of tests is performed in which every variable is set at its expected high and low values (and possibly also at intermediate values), and the effects are recorded for each set of values. The results are analysed using the technique of *analysis of variance* (ANOVA), which determines the magnitude of the effect of each variable, and of the interactions between variables. The only assumptions necessary are that the variables are normally (or approximately normally) distributed and that their effects are approximately linear over the range tested.

Statistical experimentation is widely used in fields such as process optimisation in the chemical and pharmaceutical industries, medical research and agriculture. It is also used in engineering, but not as widely as it should be. This is mainly because few engineers have been taught the method, but also because most engineering product and process designs can be analysed and tested using traditional, deterministic, theory-based methods. The main engineering applications have been for problem-solving and for optimising production processes, such as flow soldering of electronic assemblies. However, since variation exists in products, processes and environments, product and process designs can be refined and made more robust, economical and reliable by appropriate use of statistical experiments.

Note that DoE explores the effects of variables within the range covered by the experiments: it does not explore the behaviour or effects of the tails of the distributions, so the results cannot be safely extrapolated beyond the ranges tested. Therefore DoE is complementary to other test methods, and the planning of tests must include consideration of the practical aspects of the distributions of the variables, as discussed earlier.

Statistical experimental methods of optimisation in engineering design can be effective and economic. They can provide higher levels of optimisation and better understanding of the effects of variables than is possible with purely deterministic approaches, when the effects are difficult to calculate or are caused by interactions. However, as with any statistical method, they do not by themselves explain why a result occurs. A scientific or engineering explanation must always be developed so that the effects can be understood and controlled.

A statistical experiment can always, by its nature, produce results which seem to be in conflict with the physical or chemical basis of the situation. The probability of a result being statistically significant in relation to the experimental error is determined in the analysis but we must always be on the look-out for the occurrence of chance results which do not fit our knowledge of the processes being studied. That is not to say that such results would be dismissed, only that we can legitimately use our engineering knowledge to help interpret the results of a statistical experiment. The right balance must always be struck between the statistical and engineering interpretations. If a result appears highly statistically significant, then it is conversely highly unlikely that it is perverse. If the engineering interpretation clashes with the statistical result and the decision to be made based on the result is important, then it is wise to repeat the experiment, varying the plan to emphasise the effects in question.

Statistical experiments can be conducted using computer-aided design software, when the software includes the necessary facilities such as Monte Carlo simulation (Chapter 5) and statistical analysis routines. Of course there would be limitations in relation to the extent to which the software truly simulates the system and its responses to variation, but on the other hand experiments will be much less expensive, and quicker, than using hardware. Therefore initial optimisation can often usefully be performed by simulation, with hardware experiments being run to confirm and refine the results.

It is essential that careful plans are made to ensure that the experiments will provide the answers required. This is particularly important for statistical experiments owing to the fact that several trials are involved in each experiment, and this can lead to high costs. Therefore a balance must be struck between the cost of the experiment and the value to be obtained, and care must be taken to select the experiment and parameter ranges that will give the most information.

4.4.1 Taguchi method

In the early 1980s Genichi Taguchi, a Japanese engineer, developed a framework for statistical design of experiments adapted to the particular requirements of engineering design. Taguchi suggested that the design process consists of three phases: *system design, parameter design,* and *tolerance design.* In the system design phase the basic concept is determined, using theoretical knowledge and experience to calculate the basic parameters to provide the performance required. Parameter design involves refining the values so that the performance is optimised in relation to factors and variations that are not under the effective control of the designer so that the design is 'robust' in relation to these.

Tolerance design is the final stage, in which the effects of random, uncontrollable variation of manufacturing processes and environments are evaluated to determine whether the design of the product and the production processes can be further optimised, particularly in relation to the cost of the product and of the production processes. Note that the design process is explicitly considered to include the design of the production methods and their control. Parameter and tolerance design are based on Fisher's DoE techniques, but using a somewhat different method of analysis.

Taguchi separated variables into two types. *Control factors* are those variables that can be practically and economically controlled, such as controllable dimensional or electrical parameters. *Noise factors* are the variables that are difficult or expensive to control in practice, although they can be controlled in an experiment, e.g. ambient temperature or parameter variation within a tolerance range. The objective is then to determine the combination of control factor settings (design and process variables) that will make the product have the maximum 'robustness' to the expected variation in the noise factors. The measure of the robustness is the *signal-to-noise ratio*, which is analogous to the term used in control engineering.

The Taguchi method includes the use of the '*brainstorming*' approach to identify and prioritise variations and their effects. In this approach all involved in the design of the product and its production processes meet and suggest which are the likely important variables and interactions and plan the experimental framework. The team must consider all sources of variation and their likely ranges so that the most appropriate and cost-effective experiment is planned. A person who is skilled and experienced in the design and analysis of statistical experiments must be a team member, and may be the leader. It is important to create an atmosphere of trust and teamwork, and the whole team must agree with the plan once it is evolved. The brainstorming approach often leads to the identification and solution of problems even before experiments are conducted. Properly managed, the approach and the methods can be almost guaranteed to be cost-effective, since they ensure that all variables (design, production, environmental) and their possible interactions and effects are considered systematically in the planning of the integrated test programme. Therefore the approach should always be applied.

Taguchi argued that in most engineering situations few if any interactions have significant effects, so that much reduced, and therefore more economical, 'fractional' experimental designs can be applied. When necessary, subsidiary or confirmatory experiments can be performed to check whether particular interactions are important. Taguchi developed a range of such design matrices, or *orthogonal arrays*, from which the appropriate one for a particular experiment can be selected. However, the more traditional methods of DoE and ANOVA can be applied if preferred.

The Taguchi approach has been criticised for not being statistically rigorous and for under-emphasising the effects of interactions. Other critics have claimed that statistical methods have little relevance to engineering design, since all important effects of variation and interactions should be known from theory, and can therefore be analysed deterministically. However, mathematical rigour is secondary to obtaining better insight and understanding, and many engineering designs involve variations and effects that are not so well understood. The planning should take account of the extent to which theoretical and other knowledge, for example experience, can be used to generate a more cost-effective experiment. For example, theory and experience can often indicate when interactions are likely or significant.

Taguchi's greatest contribution has been to foster a much wider awareness of the power of statistical experiments for product and process design optimisation and problem-solving. The other major benefit has been the fostering of the need for an integrated approach to the design and test of the product and of the production processes. It is important to appreciate that Taguchi has developed a method that deliberately economises on the number of trials to be performed in order to reduce experiment costs, and this is of course an important consideration in engineering.

We will discuss DoE and the Taguchi approach to testing further in Chapter 6. The methods are described in detail in References 4, 5 and 6, and in other books.

4.5   DISCRETE VARIATION

Values or properties that can vary only over a finite number of states, for example a coin that can be tossed to show heads or tails, a switch that can be either on or off, or a missile that either works or fails, are *discrete* variables. The probabilities of outcomes are sometimes determined by the nature of the system, for example heads or tails with coin tossing, 1–6 with dice, and a number on a roulette wheel. For other systems, like switches and missiles, the outcomes are uncertain and estimates of future behaviour cannot be made without recourse to other information, including data on past performance. For example, if eight missiles work out of 10 fired, we could say that the reliability of the next batch of 10, of the whole population, or of future production, will be about 80 percent. However, this prediction is not as certain as the probability of throwing a 6 with a die, since many other factors could influence the reliability, such as design and process changes, redefinition of failure criteria, changes in manufacturing quality, etc.

In addition, the element of chance (or luck) affects the outcomes. If we throw three 6's in a row, it should not influence our judgement about the probability of having a 6 on the next (it will of course be unchanged, 1/6). However, were the eight missile successes truly representative of the population reliability, or were we 'lucky/unlucky' on those tests? If we know the causes of the failures we might be able to make judgements of and influence future reliability. Therefore our interpretation of the results and the way we use them to make predictions must be tempered with engineering judgement.

Statistical models that describe these situations are the *binomial* distribution (for situations in which there can be only one of two possible outcomes) and the *Poisson* distribution (for any discrete number of outcomes).

4.6 CONFIDENCE AND SIGNIFICANCE

The interpretation of data from statistical tests always involves the questions of statistical confidence and statistical significance. We will refer to these as *s-confidence* and *s-significance*, to distinguish them from 'practical' or engineering interpretations. S-confidence is the mathematically derived likelihood that the result obtained from a sample will represent the behaviour of the population from which it is (randomly) drawn. S-confidence is usually expressed as *confidence limits*. For example, if 10 items out of a sample of 10 work, the 80 percent s-confidence limits of the population reliability will lie between 85 percent (lower s-confidence level) and 98 percent (upper s-confidence level). If the sample size is greater, the s-confidence limits would be narrower, and vice versa.

S-significance relates to the situation when we want to know the likelihood that a sample result indicates deviation from a population parameter, or that two results indicate that they are derived from different populations. For example, if a process shows different variations with two operators, does the difference arise because of the different operators, or is it due to chance? Mathematical tests can be applied to the data to determine the s-significance of the hypothesis that the operators are different.

However, we must always bear in mind that the statistical data is seldom the only information available. For example, if the test described above involved releasing a weight to see whether it drops, and it dropped each time, we could make a statement about the s-confidence limits on its 'reliability'. However, as engineers and scientists we know that it will drop every time, even if we perform no tests, because we know about gravity, so our *practical* confidence in 100 percent reliability is 100 percent, irrespective of the number of tests. If the problem is different, say the effect of friction on the release mechanism, then we should apply our knowledge of friction and lubrication, rather than just perform a series of drop tests followed by statistical analysis.

The statistics provide clues. They do not provide explanations. Statistical results should be taken into account, but engineering knowledge must always provide the ultimate basis for decisions. The use of statistical confidence and significance in engineering applications is seldom useful, and is often misleading.

4.7   RELIABILITY

The 'textbook' definition of *reliability* is the probability that an item will not fail during a test or over a period of time, under defined operating conditions. Reliability is also expressed as the *mean time to failure* (MTTF) (for an item that can fail only once, like a light bulb, microprocessor or interplanetary spacecraft), or the *mean time between failures* (MTBF), for an item that can fail and be repaired a number of times, like a car or a circuit assembly. The MTBF measure implies that the rate at which failures occurs is, on average, constant over the time considered, and then MTBF is equal to the inverse of the *failure rate*. The failure rate is most often used at the level of components, and is often expressed as failures per million hours (fpmh), and denoted as λ (lambda).

Reliability data is published for many component types, most notably US MIL-HDBK-217 (Reliability Prediction for Electronic Systems), described in Chapter 3. Sources such as these are used for 'predicting' the reliability of new designs, by adding the failure rate contributions of all of the components. However, this approach is fundamentally wrong and misleading, on several grounds. The most notable, from the point of view of test engineering, are the following.

- Engineering components have no intrinsic property of 'failure rate'. As discussed in Chapters 2 and 3, failures have causes which are mostly extrinsic to the components.

- The mathematical 'models' used to represent the effects of application conditions (temperature, etc.) on failure rate are not consistent with the physical and other causes of failure.

- The 'data' in such sources is nearly always of very dubious quality and several years out of date in relation to technology and capability, thus giving very pessimistic predictions in comparison with what can be achieved with adequate engineering.

- Failures do not generally occur at constant average rates over time. They can be time-dependent, cyclic stress dependent, etc.

- Component failures do not necessarily cause system failures, and system failures are not always the result of component failures.

Methods have also been developed for 'demonstrating' reliability, notably US MIL-STD-781, based upon measuring the MTBF (total time on test/total failures) in planned tests, and comparing it with pass/fail criteria based on the s-confidence limits. These methods are also inappropriate and misleading. They are described and discussed in Chapter 12.

The quantitative approaches that have been used and taught for reliability 'prediction' and 'demonstration' are controversial, and should be avoided. Reliability is the absence of failures, and failures have many causes (physical, chemical, processes, human), effects, and patterns of occurrence. Testing in development is one way of driving reliability improvement, by indicating opportunities for product and process redesign. Testing in manufacture contributes by preventing defective items from being delivered. Testing should not be used to generate reliability statistics or to comply with pass/fail reliability criteria.

4.8 SUMMARY

Variation in engineering is usually more complex and difficult to deal with than most 'natural' variation, for the following reasons.

- It seldom follows the conventionally taught mathematical form of the normal distribution.

- In many situations the parts of the distributions of most concern are the extreme values in the 'tails'. This is where the data is always less frequent and more uncertain, and where conventional statistical methods are most misleading.

- It can change over time (cycles, distance, batch number, etc.).

- Interaction effects can be difficult to predict and understand.

- Much of it is caused by human behaviour (skill, understanding, etc.).

- Most engineering teaching covers no more than conventional statistics, and engineers therefore tend to be uncertain about how to deal with variation and sceptical about the application of statistical methods.

Therefore we must take full account of these realities. In the integrated engineering approach to new product design, test and manufacture we must ensure that, as far as practicable, all variations that can affect performance, reliability, durability and costs are identified, understood and controlled, using the practical methods discussed in this and other chapters.

Reliability 'textbook' teaching and standards have also over-emphasised inappropriate quantitative and statistical approaches that have been shown to be ineffective and misleading. High reliability (and durability) are achieved by the application of good engineering in the widest sense, to design, manufacture, and maintenance (when appropriate). Testing is a very important aspect of this. Statistical methods can be used to help to plan tests and to interpret the results, but they must be applied with proper appreciation of their limitations, and in ways that are consistent with the practical engineering realities.

All of the topics discussed in this chapter are described in more detail in Reference 3. There are many good books on applied statistics, such as References 5–10. Methods for analysing data from tests, including statistical methods, are described in Chapter 12.

REFERENCES

1. Shewhart, W. A., 1931, *The Economic Control of Quality of Manufactured Product*, Van Nostrand.
2. Deming, W. E., 1986, *Out of the Crisis*, MIT Press.
3. Pyzdek, T., 2001, *The Six-sigma Handbook*, McGraw-Hill.
4. O'Connor, P. D. T., 1995, *Practical Reliability Engineering* (3rd edn), John Wiley & Sons, Ltd.
5. Hines, W. W. and Montgomery, D. C., 1980, *Probability and Statistics in Engineering and Management Science*, John Wiley & Sons, Ltd.
6. Park, S. H., 1996, *Robust Design and Analysis for Quality Engineering*, Chapman & Hall.
7. Moroney, M. J., 1965, *Facts from Figures*, Penguin.
8. Wheeler, D., 1993, *Understanding Variation: the Key to Managing Chaos*, SPC Press.
9. Breyfogle, F. W., 1992, *Statistical Methods for Testing, Development and Manufacturing*, Wiley-Interscience.
10. Nelson, W. A., 1990, *Accelerated Testing: Statistical Models, Test Plans and Data Analysis*, John Wiley & Sons, Ltd.

5

Design Analysis

5.1 INTRODUCTION

Testing is not the only way that we can obtain knowledge about the capability of an engineering design to achieve its requirements. All designs are based upon knowledge of the relationships and parameters involved, so, in principle, we should be able to create designs and analyse them to obtain confidence in their capabilities in terms of performance, reliability, safety and durability. Getting designs right the first time by applying knowledge and experience is obviously the quickest and cheapest way. For example, a simple structure or circuit can usually be designed by one engineer using only basic knowledge and manual calculations, but the design of a complex structure subjected to dynamic loads or a complex logic circuit will require the application of more powerful computer-based analytical methods.

Manual and computer-driven analyses are usually cheaper than testing, and testing also adds to project timescales. Therefore, designs must always be analysed as far as is practicable, to obviate or to reduce the need for testing. Analysis does not always remove the need for test, since it is usually prudent to check or confirm the results of analyses by test. Analysis can also show up which aspects of the design might need to be further investigated or confirmed by testing, by highlighting aspects that present risks or uncertainty.

For some designs it is too expensive or impracticable to perform tests. The Space Shuttle re-entry manoeuvres could be tested only by using small-scale models in high-speed wind tunnels, and the first orbital flight was the first time that the full-scale system had to perform over the full operating envelope. In the event the controls proved to be barely adequate to prevent the craft from pitching up on re-entry and being destroyed, though the simulations and wind tunnel tests had indicated that there would be an adequate margin. Likewise, large structures such as buildings and bridges are not usually tested before being put into use.

Reliability, durability, and performance aspects that can be affected by variation are difficult to analyse effectively, so further testing is nearly always necessary to confirm these aspects. Testing for reliability, durability and variation will be covered in the next chapter.

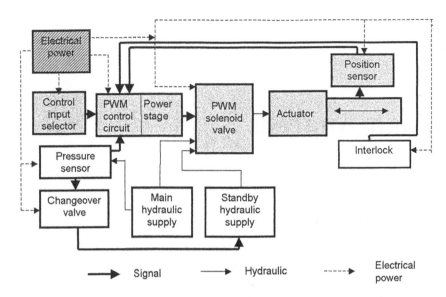

Figure 5.1 Electrohydraulic actuator system conceptual design

Analysis and testing must be managed integrally in order to obtain the greatest benefits from both. Designers must have knowledge not only of the relevant theory and technology, but also of the appropriate analysis and test methods. They must be aware of the capabilities of the analysis methods and, equally importantly, of their limitations.

In this chapter we will review the main methods and software that are used for analysing engineering designs. The use of appropriate software can greatly ease the engineering design task, and often makes possible analyses that would otherwise not be feasible.

The system shown in outline in Figure 5.1[1] will be used as an example for several of the analyses. The basic system requirements are to provide a motion control system for a safety-critical application with specified performance in terms of thrust, frequency response, etc. The system must not fail with the actuator extended. The initial concept design solution is as follows.

- A pulse-width modulated (PWM) solenoid valve provides the supply to a hydraulic actuator. The inputs to the solenoid valve are generated by an input selector and a feedback control circuit and power output stage. (Possible alternative: continuous electrohydraulic servo valve controlled by analogue feedback signal.)

---

[1] Shaded blocks based on a design provided courtesy of The MathWorks, Inc. For the full system description and analysis contact The MathWorks, Inc.

- There are two hydraulic power supplies: a main supply and a standby supply. The standby supply is from a pressurised reservoir which allows system operation long enough for the actuator to be retracted. Failure of the main hydraulic supply is detected by the pressure sensor, which sends a signal to the control circuit and to a solenoid-operated changeover valve which activates the standby supply.

- There is one electrical power supply. (Possible alternative: backup supply.)

- There is a safety interlock which must be activated before the actuator can operate, and which retracts the actuator and shuts the system down if the actuator extends more than a specified distance.

5.2   QUALITY FUNCTION DEPLOYMENT

Quality Function Deployment (QFD) is a technique that was originally developed and applied in Japan to identify and prioritise all of the 'quality' requirements of a new product, and to relate them to the product features that influence their achievement. 'Quality' in this context refers to any characteristic that might positively influence customer preference for the product, and therefore QFD is appropriate to all aspects, including performance, reliability, appearance, cost, etc. QFD is not an analysis in the sense of determining how the design will work: instead it is a review method to determine and prioritise the features and methods that are necessary to achieve the optimum product, in a way that is consistent and open. It can therefore provide an excellent basis for focusing attention on how requirements and methods might interact or conflict, and for identifying and planning the design analysis and testing work, including product features and methods that should be considered in the test programme. For maximum benefit it should be performed at or soon after the inception of the project.

A matrix format is used, as illustrated in Figure 5.2. This shows a partial QFD for the conceptual design of the system in Figure 5.1. The requirements are listed in the rows and are given importance ratings. A scale of 0 (unimportant) to 5 (essential or critical) can be used. The features of the design that will enable the requirements to be achieved are listed in the columns. The extent to which each feature influences each requirement is then indicated in the matrix, using the same scoring principle. The total score for each feature is the sum of the products of requirement importance times the influence score.

The QFD chart enables other aspects to be highlighted for attention. Features that interact can be indicated, as shown. The extent to which competitors' products achieve the requirements can be indicated in 'benchmark' columns, to assist in determining importance ratings.

The QFD method is 'deployed' downwards, so that, for example, the lower-level assemblies and components and the manufacturing processes are analysed

Design analysis

REQUIREMENTS	UNITS	IMPORTANCE	1 Selector	2 PWM controller	3 Solenoid valve	4 Actuator	5 Position sensor	6 Interlock	7 Pressure switch	8 Changeover valve	COMMENTS	BENCHMARK
INTERACTIONS			2	1,3	2	3	2,4	2	2	7		
1 Response rate	15 Hz	5	0	4	4	4	2	0	0	0	Analyse, test	4
2 Max. thrust (hyd. pressure 200 bar)	200 N	5	0	0	0	5	0	0	0	0		2
3 Accuracy	0.5 mm	4	2	2	4	2	5	0	0	0	Analyse, test	3
4 Max.overshoot	10 mm	2	0	0	3	2	2	5	0	0	Analyse, test	2
5 Fail safe		5	2	4	3	4	2	5	5	5	Analyse, test	3
6 Hyd. supply changeover	150 bar	4	0	0	0	0	0	0	5	5	Hydraulics test	1
7 Max. temp	45°C	4	2	3	2	2	3	0	2	2	HALT	1
8 Min. temp	-20°C	4	2	4	2	2	4	0	2	2	Environmental test	1
9 Max. weight	12 Kg	3	3	1	3	4	2	1	2	2		2
10 Reliability	Very high	5	3	3	4	3	4	2	3	3	HALT	3
11 Durability	20 years	4	4	2	4	4	4	2	3	3	HALT	3
12 Max. production cost	$400	4	3	2	4	4	2	1	2	3		4
SCORE			86	110	134	152	122	70	102	106		
NOTES			1	1								

Notes: 1. Consider/analyse continuous servo control (accuracy, cost)

Figure 5.2   QFD of electrohydraulic motion system (top level)

in the same way, so that every component and process is optimised in relation to the top-level requirements and their importance ratings. Particular requirements and features can be assessed in separate QFDs, for example the causes and effects of variations and the methods to control them.

The best way to execute the QFD is to set up a project team, comprising the key people who will be responsible for design, development, manufacture, marketing, etc., and to 'brainstorm' the identification of requirements, features, scores, etc. It can also be helpful to include potential customers and suppliers of important items.

The payoffs of performing QFD are that:

- It forces an integrated, 'concurrent' approach to engineering design, development, manufacture and support.

- It forces understanding of customer requirements and priorities at all levels and phases (*'voice of the customer'*).

- It forces consideration of alternatives.

- It reduces changes and unpleasant surprises

- It forces rational approaches to manufacturing controls and tests.

- It generates a better, cheaper product, faster and at less cost.

QFD software is available. Alternatively, a spreadsheet can be created and tailored to the application. An advantage of using a customised spreadsheet for this work is that it can also be used to perform calculations, including ANOVA if the results of statistical experiments are added.

QFD is described in detail in Reference 1.

5.3   DESIGN ANALYSIS METHODS

5.3.1   Mathematical

The main mathematical methods for analysing engineering designs are:

- Functional calculations based upon the relationships involved (stress, strength and strain; electrical potential, resistance and current; thermodynamics; control laws; logic; etc.).

- Mathematics software which can be used to create and solve complex equations and present graphical outputs. References 2 and 3 describe mechanical, and References 4 and 5 describe electronics and control applications, respectively.

These methods can be used for analysing simple to moderately complex designs, such as of static mechanical components, simple mechanisms and small systems. Typical capabilities of mathematics software include:

- General mathematics equation solving
- Graphical outputs
- Data acquisition
- Model libraries for components, etc.
- Specialised modules for applications such as signal processing, image processing, communication systems, statistics, fuzzy logic, neural networks, C/C++ compilers, web and spreadsheet interfaces, report generators, etc.

Figures 5.3 and 5.4 show examples of the results of analysis of the system in Figure 5.1. More specialised methods are available for analysis of more specialised or complex designs, and these will be described later.

5.3.2   Mechanical

Mechanical components and system designs can be created and analysed using *computer-aided design* (or drafting) (CAD) software. Analysis capability includes dimensional tolerance effects, clearances, paths of moving parts, tool accessibility, etc. Most modern CAD software enables three-dimensional (3-D) views to be created. CAD software is usually the starting point for many other *computer-aided engineering* (CAE) tasks, such as stress and vibration analysis, creation of

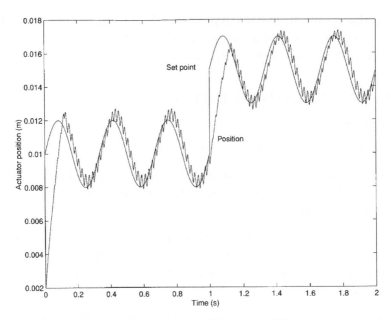

Figure 5.3 Simulated piston motion: Mathcad™ Simulink™ analysis (courtesy The Math-Works, Inc.)

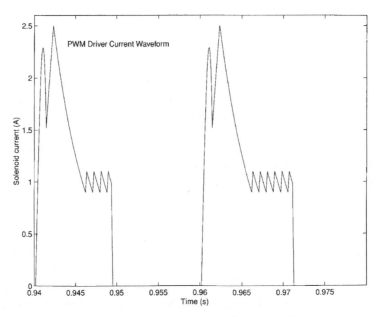

Figure 5.4 Simulated solenoid current: Mathcad™ Simulink™ analysis (courtesy The MathWorks, Inc.)

production machining and measurement (CMM) instructions, and virtual reality and rapid prototyping inputs (see later).

Finite element analysis

Finite element analysis (FEA) is a generic computer-based technique which successively evaluates the effects of disturbances (mechanical stress, vibration, temperature, etc.) on a design. The design to be analysed is converted from its original, continuous form to a finite element approximation by sampling values at discrete intervals, defined by a *mesh*, and the calculations are performed sequentially from element to element throughout the mesh. FEA methods are applied to study mechanical stress patterns in static or dynamically loaded structures, strain, responses to vibration or *nodal analysis* (resonant frequencies, nodes and displacements), heat flow, and fluid flow. FEA programs interface directly with CAD software, which automates the mesh generation. FEA can be integrated with material properties and loading pattern data to perform fatigue life analysis. Reference 6 describes these applications.

Figure 5.5 shows an example of FEA application to the design of the mounting bracket for the actuator in the example of Figure 5.1, and Figure 5.6 shows the results of fatigue analysis. The stress contours in Fig. 5.5 are based upon the application of a static axial load, as shown. In principle, the fatigue endurance contours would follow the same pattern, if no allowance was made for other factors that can influence them, such as surface finish, material property differences, etc. Also, for more complex cyclic loading situations, including multi-axial loads, the fatigue endurance profiles would take account of cycle counts at the different stress levels and principal directions, and could therefore be very different to the FEA stress profiles. Software is available to perform this analysis.

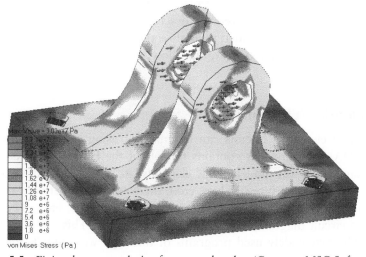

Figure 5.5 Finite element analysis of actuator bracket (Courtesy MSC Software)

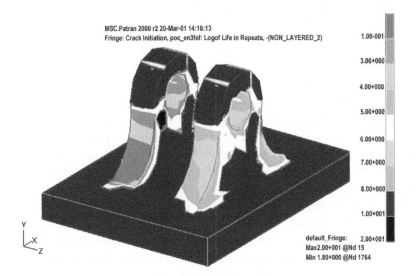

Figure 5.6 Fatigue analysis of actuator bracket performed using nCode fatigue software (courtesy MSC Software)

The use of FEA for studies of fluid flow, including aerodynamics, is called *computational fluid dynamics* (CFD). CFD software can be used to analyse flow of liquids and gases, in streamlined, turbulent, viscous or supersonic flow, as well as heat flow and combustion. Typical applications include aerospace and automotive design, engine design (inlet and exhaust flows, combustion), heating and ventilation systems, aerodynamic effects of structures, and chemical process flows. CFD has to a large extent replaced the need for expensive wind tunnel testing for aerodynamics. Reference 7 describes CFD.

FEA and CFD can be extended to analyse the simultaneous effects of different physical phenomena, such as heat, fluid flow, electromagnetic fields, etc.

5.3.3 Electrical/Electronic

Electrical and electronic circuits can be analysed using *electronic design automation* (EDA) software that contains models of the behaviour of components (Ohm's Law, transistor models, logic elements, etc.). The input data consists of the component details (types, values, etc.), the connections between the components, and the electrical input values (power, waveforms, etc.). The software creates the circuit diagram and analyses its behaviour. This is called *schematic capture*. The first widely used programs of this type were SPICE for analogue circuits and HiLo for digital, and proprietary versions of these are available, ranging in capability from relatively small-scale programs to very advanced software. For large digital circuits the *VLSI Hardware Description Language* (VHDL) is used.

Modern EDA programs integrate many capabilities. These typically include:

- Models for most of the commercially available standard components

- Schematic capture, circuit diagrams and layouts

- Analogue, digital and mixed-signal circuits (SPICE, VHDL)

- Important aspects of circuit performance, such as responses to power and data inputs, accuracy, frequency response, timing, etc., can all be 'tested' using 'virtual probes' and graphically displayed on 'virtual instruments' before hardware is built.

- Sensitivity analysis: the software identifies parameters to which circuit behaviour is most sensitive.

- Stress analysis: the software analyses voltage, current and power stress levels, both transient and steady-state, and compares them to rated stresses.

- Tolerance analysis: parameter tolerances and other variation distribution assumptions can be input, and the software will automatically analyse circuit behaviour in response to these and identify the most important variables.

- Production data such as instructions for bare circuit card conductor routing and component placement

- Testability analysis (see Chapter 8)

- The effects of potential failures can be analysed by inputting appropriate parameter values or other fault conditions, such as infinite resistance values for open circuits.

Some EDA programs include further capabilities, such as:

- Microwave frequency analysis

- Multi-technology simulation (electronic, mechanical, fluid, etc.)

- Waveform analysis (analogue and digital), including Fourier, Bode, Nyquist, etc.

- Thermal analysis (Figure 5.7)

- Inputs from other EDA programs

- User-definable component models.

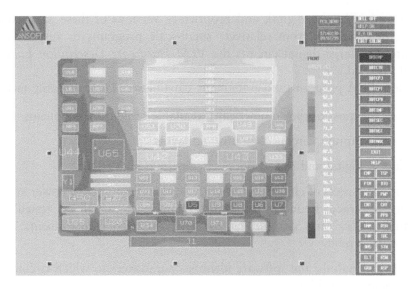

Figure 5.7   Thermal analysis of a printed circuit board design (courtesy Ansoft Corp.)

Electromagnetic effects

Electromagnetic effects can be analysed using mathematics software or specialised programs. These include capabilities for two- and three-dimensional mapping of electrical and magnetic fields, as well as resulting mechanical aspects such as force and torque. The software is based on Maxwell Law mathematics solving and FEA. Designs of items such as high-speed circuits, electric motors, solenoids, actuators, loudspeakers, antennas, etc., can be simulated and optimised. Typical modern software includes capabilities such as:

- Automatic mesh generation for FEA
- High-frequency circuit performance (RF, microwave, fast digital)
- EMI/EMC analysis (external and internal: delays, crosstalk, etc.)
- Integrated circuit performance, on-chip and package layout aspects
- Analysis of other components, such as connectors, enclosures, etc.
- Links to other EDA programs
- Thermal analysis.

Reference 8 describes some computerised analysis methods for electromagnetics. Figure 5.8 shows an example of signal integrity analysis of a circuit board design, and Figure 5.9 shows an electromagnetic field analysis.

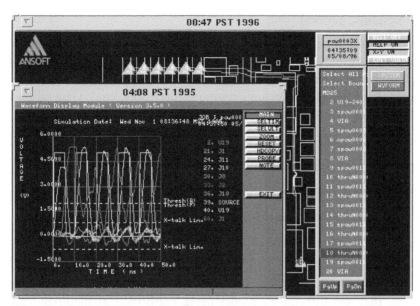

Figure 5.8   Signal integrity analysis of a circuit board design (courtesy Ansoft Corp.). (The square waveforms are the input pulse, and the distorted ones are the same pulse after propagating through board-level interconnects. The low-amplitude waveforms are crosstalk signals on nearby nets)

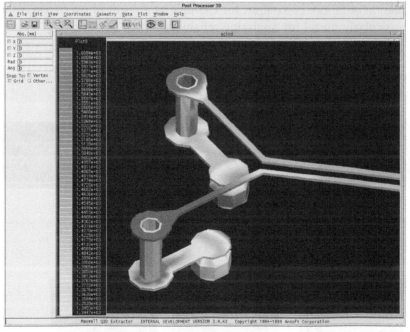

Figure 5.9   Electromagnetic field strength analysis (courtesy Ansoft Corp.). (AC current density near some signal vias and solder balls in a BGA package, as computed by Maxwell SI 3D)

5.3.4 Systems, general

System simulation methods enable systems that include complex controls, logic, data, feedback, displays, etc. to be designed, 'tested' and optimised. Some of the available software is suitable for a wide range of applications, while others are more specialised (signal processing, computer networks, etc.). General-purpose simulators include features such as:

- Control theory and signal processing problem solving, waveform analysis, etc.

- Graphical inputs and outputs

- Parameter sensitivity and variation analysis

- Automatic generation of control software

- Designs that involve multiple technologies (mechanical, electrical/electronic, software) can be analysed using appropriate software which includes models of non-electronic components.

- Features that are difficult to model using mathematical methods, such as deadband, backlash, hysteresis, etc., can be included and their effects analysed.

- The user can also create component models, all in the parametric units of the relevant technologies.

- Simulation of failure behaviour, for example as a result of overstress, so it can be particularly useful for analysing reliability.

- Monte Carlo modelling (see later).

Reference 9 describes methods and applications. Figure 5.10 shows an example.

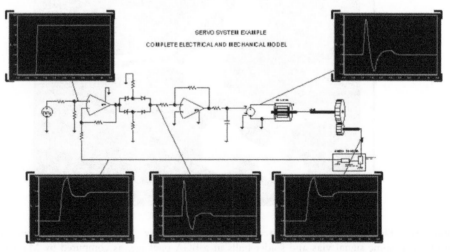

Figure 5.10 Simulation of multi-technology design (courtesy Avanti Corp.)

5.3.5 Monte Carlo simulation

Monte Carlo simulation is a technique for simulating the operation of any system which operates over a time frame, and which includes multiple parameters, variables and interactions. In principle, any such system can be modelled analytically by development and solution of the relevant equations. However, for real systems this can become an impracticable task, since the complexity of the resulting equations increases exponentially with the increase in parameters, variables and interactions. In practice we often make simplifying assumptions when analysing the behaviour of complex systems in order to make the analysis tractable, but such simplifications can be unrealistic. (The kinds of problems set in engineering examinations are often unrealistically simple, in order to ensure that they can be 'solved' in a short time. Real engineering problems are seldom like that.) For example, if we want to determine the likely distribution of fatigue life consumption on a vehicle component, when the vehicles are introduced to service over a period, are subjected to different applications, environments and rates of use, and some components are replaced for fatigue failures or other causes, an analytical approach would be difficult and unrealistic. However, with the Monte Carlo approach the problem is simulated by defining the system logic, then assigning parameters and distributions or probabilities. The software then models the probabilistic aspects by sampling from random numbers, whose ranges or values determine outcomes such as, in this case, utilisation and replacements. The sampling process is repeated a large number of times, so that the range of system possibilities is simulated. The outputs can be presented graphically, so that they show the range of likely outcomes and sensitivity to input assumptions, and they can be analysed statistically. The Monte Carlo approach enables us to simulate situations that are too complex for analytical treatment, and the processing task increases only linearly with added inputs. The method is also much easier to understand than analytical formulae.

Monte Carlo simulation is widely used for analysing complex, multi-variable systems such as production and repair lines, computer networks, transport fleet operation, and logistics. Such programs include facilities for statistical model inputs, graphics outputs, processes, animations, etc.

The major EDA and system simulation programs include Monte Carlo capabilities. These can be used to perform tolerance analyses and manufacturing test yield analyses on system, circuit and component designs.

Reference 10 is a good introduction to Monte Carlo simulation.

5.3.6 Other system simulation methods

Hardware in the loop simulation is the term used to describe computerised analysis of a system design, but with particular hardware units being used in place of software models. For example, the pneumatic actuator system design could be analysed with simulated selector, servo valve and actuator, but using a real control circuit, or vice versa. This approach can be useful and valuable when hardware units are available, as it saves effort on generating software models and it can produce results that are more realistic and credible than using entirely software simulations. It can be applied at later stages of the project, as actual hardware becomes available for test.

Virtual reality (VR) software can be used to add dynamic effects and other features to designs, such as viewing from different perspectives, showing assembly or dismantling processes, etc. VR is often used for aesthetics design and marketing applications, but it can also provide valuable support to engineering development, for example to check assembly and maintenance aspects.

Rapid prototyping (RP) is a technique for creating spatial models of designs using the CAE drafting outputs. RP models can be used for easy realisation of how components and assemblies fit together, interferences, etc.

5.4   ANALYSIS METHODS FOR RELIABILITY AND SAFETY

In addition to considering all the factors that will enable a design to work correctly, we should attempt to identify all the ways by which it might fail. Chapters 2 and 3 described these in general terms. As explained, the analytical methods described above have only very limited capabilities for this kind of work. Many failures occur across interfaces, or involve phenomena that are outside the scope of the computer models and methods. For example, FEA takes no account of corrosion, and most EDA circuit models assume that components work correctly. Analysis methods have been developed to help to identify failure causes and effects. The most important of these are described below, and in more detail in References 11 and 12.

5.4.1   Load and strength analysis

All components subjected to loads that could cause them to fail should be analysed to ensure that adequate margins exist. The effects of load and strength variation, combined stresses and time-dependent weakening (fatigue, wear, etc.) must be included. In Chapter 4 we discussed the design situation in which variable loads must be resisted by components of variable strength. The *load–strength analysis* (LSA) should include the following:

- Determine the most likely patterns of variation of load and strength
- Evaluate the safety margin for intrinsic reliability
- Determine protection methods (load limit, derating, screening, other quality control methods)
- Identify and analyse strength degradation modes

Item (material, function)	Load (worst case, combined)	Frequency/ duration probability	Data source	Combined effect	Strength	Safety Margin	Degradation	Controls
Actuator bracket	1. 100 N	Continuous	Spec.		400 N	4		FEA
	2. Cyclic ±100 N	50 Hz, 1M cycles	Spec.	Fatigue	?	?	Fatigue	Analyse, test
PWM					Max Rating		Insulation	Test
Solenoid:	1. 2.5 A	1. 50 Hz	Analysis	150 W	200 W,	Power OK	breakdown?	Review selection?
1. Coil current	2. 45°C	2. Continuous	Spec.	90°C	85°C	Temperature		
2. Temperature ambient						?		

Figure 5.11 Load–strength analysis of motion system (partial)

- Test to failure to corroborate, analyse results

- Correct or control (redesign, safe life, quality control, maintenance, etc.).

Figure 5.11 shows a format that could be used for documenting the LSA.

5.4.2 Failure modes and effects analysis

Failure Modes and Effects Analysis (FMEA), also called *Failure Modes, Effects and Criticality Analysis* (FMECA), is a method for tabulating all of the components (or functions) within a design, and asking the following questions of each one:

1. How can it fail in the application? (failure mode)

2. How likely is each failure mode? (failure probability or failure rate)

3. What would be the effect of each failure mode? (failure effect)

4. How critical is the effect? (failure effect criticality)

5. What should or could be done about it?

It can be helpful to use pre-assigned codes for effects (e.g. $0 =$ no effect, $1 =$ fail unsafe, $2 =$ fail safe, etc.), and judgement values, for example a 0–1 decimal scale, for likelihood and criticality ($0 =$ impossible/not at all critical, $1 =$ extremely likely/highly critical).

Supplementary questions can also be put and the answers tabulated. These could include:

1. Warning or indication of failure mode

2. Pattern of failure, e.g. wearout

3. Effects on safety, cost, etc.

4. Detectability by test equipment or built-in-test

5. Repair methods.

An example of part of a FMEA on the motion system design is shown in Figure 5.12.

The prime objective in performing a FMEA is to identify how failures might occur, so that, as far as is practicable, the causes can be eliminated or reduced, or the effects mitigated. There can also be secondary objectives, including optimisation of test and maintenance. FMEA is, therefore, a method to reduce the 'uncertainty gap' in relation to reliability before designs are finalised and hardware is made and tested. If we believe that the engineers involved are incapable of creating errors or weaknesses in the design, then the prime objective

Item	Failure modes	Failure likelihood	Failure effect	Effect criticality	Remarks
Electrical power	No supply	0.2	System fails (unsafe)	1	Provide standby power?
Main hydr. supply	No supply	0.4	Revert to standby	0.5	
Standby hydr. supply	No supply	0.2	nil (IF MAIN ok)	0.4	Checked on maintenance
Pressure sensor	Failre to detect	0.2	NIL (if main supply	0.4	Checked on maintenance
Changeover valve	Failre to change over	0.1	NIL (if main supply OK)	0.4	Checked on maintenance
Control circuit	No output	0.1	System fails (unsafe)	1	Circuit FMEA HALT
Power stage	No output	0.2	System fails (unsafe)	1	Circuit FMEA HALT
Solenoid valve	Stuck open	0.1	System fails (unsafe)	1	HALT
	Stuck closed	0.2	System fails (safe)	0.4	
Position sensor	No output	0.1	System fails (unsafe)	1	HALT
Actuator	Stuck (Extended)	0.1	System failes (unsafe)	1	
	Stuck (retracted)	0.1	System fails (Safe)	0.5	

Figure 5.12   FMEA of motion system design (partial)

is inapplicable. However, most engineering design is too difficult to enable us to justify such assumptions. In practice, if one or more failure modes are detected that result in design improvements, the cost and effort of the analysis is usually repaid.

FMEA should be performed by a person or a small team, who can work independently of the designers. Engineering design is a creative, problem-solving task, involving a range of specialist knowledge. FMEA is a more deductive task, for which the analyst needs to understand the design solutions, and the interfaces between all of the specialist aspects. Ideally there should be close teamwork between the designers and the engineers performing the FMEA. The results of the analysis must then be considered and acted upon, so that the design can be improved.

FMEA is not a trivial task, and can involve many hours or weeks of work. It can also be difficult to trace the effects of low-level failures correctly through complex systems. EDA programs can be used to assist in the analysis, thus aiding the task of working out the effects of component-level failures on the operation of complex systems which have been developed on them. Even with aids such as these, FMEA can be an inappropriate method for some designs, such as digital electronic systems in which low-level (e.g. transistor) failures are very, but uniformly, unlikely, and the effects are dynamic in the sense that they could differ widely depending upon the state if the system. FMEA is not appropriate for software designs.

One way of escaping from the task of identifying the failure modes of many components in large systems, or when project time constraints place limits on the depth of the analysis, is to change the starting point of the analysis to the question: '*how can the system fail?*'. We then look at the outputs, and work downwards through the system to identify the most likely causes.

FMEA is widely used in many industries, particularly in those for which failures can have serious consequences, such as military, aerospace, automotive, medical equipment, etc. Some industries have established standardised approaches (the US Military Standard is MIL-Handbook-1629, and the US automotive companies have also produced a guidance document). However, these present rather rigid approaches, which furthermore are not appropriate for systems involving modern digital electronics, so they should be used only to the extent required by contracts and as valid for the technology.

5.4.3   Sneak analysis

Sneak analysis (SA) is a method to detect ways in which a system could fail, even though all components are working correctly. It was developed in response to the fire in the Apollo capsule, which resulted in the deaths of the three astronauts before the first NASA mission to the Moon. This was caused by a sequence of switch settings which caused an unprotected electrical short-circuit to earth: the system had been inadvertently designed to catch fire, and the FMEA could not have detected this.

5.4.4 Fault tree analysis

Fault Tree Analysis (FTA) is a method for analysing the effects of single or multiple simultaneous failures on system operation. It is based upon creating a logical representation of the system in terms of failures and their effects. A separate FTA is created for each defined '*top event*', or failure effect that is to be prevented or minimised. Figure 5.13 shows a partial example, based on the system design as analysed in Figure 5.12. FTA is used for analysing most designs that have severe safety requirements, such as controls for safety-critical processes, aircraft controls, railway signalling, etc. Software is available for creation of the 'tree' and for determining the lower-level failure combinations that could cause the top event and evaluating the probability of its occurrence, based upon the input logic and lower-level failure rates or probabilities.

5.4.5 Hazops

Hazard and operability study (HAZOPS) is a technique for the systematic determination of the potential hazards that could be generated by a system,

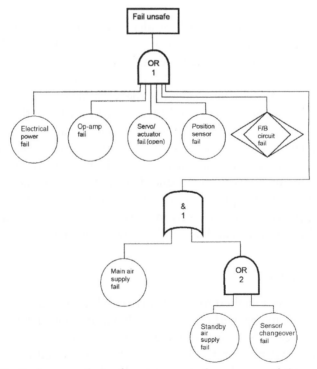

Figure 5.13   Fault tree analysis of motion system for top event 'fail unsafe' (partial)

Component/ function	Failure/ deviation	Possible cause(s)	Consequences/ event no.	Safeguards	Action
Electrical power	No power	1. Main power fail	System failure. (1)	Provide standby power?	System design
	2. Connector				
Hydraulic supply	Main AND standby fail	1. Main AND pressure sensor fail	System failure. (2)	Checks on maintenance	Maintenance schedule
		2. Main AND changeover valve fail	System failure. (2)		
PWM circuit	Permanent 'on'	see FMEA	System failure. (3)	To be determined	Analysis, test
Solenoid valve	Stuck open	Corrosion	System failure. (3)	To be determined	Test

Figure 5.14   HAZOPS on motion system (partial)

and of the methods that should be applied to remove or minimise them. It is used in the development of systems such as petrochemical plant, railway systems, etc., and usually is part of the mandatory safety approval process. Figure 5.14 shows an example of the format used.

5.5   DESIGN ANALYSIS FOR PROCESSES

The processes that will be used to manufacture and maintain the product must be understood and optimised. It is essential that all of the processes are capable of being performed correctly and efficiently, whether by people or by machines. Therefore the designers must know the methods that will be used and their capabilities and limitations, and must design both the product and the processes accordingly. The test programme must include tests of all of the processes that have been shown by the analyses to be critical or important.

The applicable analysis methods that can be used are described below.

5.5.1   Process FMEA

A *process FMEA* (PFMEA) is performed as described above for FMEA, but instead of asking 'how can the component or function fail?', we ask 'how can the process fail?'. Each process task is considered in turn, with the objective of identifying potential problems so that improved methods or controls can be set up.

5.5.2 'Poka yoke'

Poka yoke is the Japanese expression for 'mistake proofing'. It is a design approach that considers the ways in which processes might be incorrectly performed, and then making it difficult or impossible to do so. Examples are templates to ensure that directional components cannot be connected the wrong way round.

5.5.3 Testability analysis

Electronic circuits and systems must be tested after assembly to ensure that they are correct, and to indicate the sources of failure. Testability and testability analysis are described in Chapter 8.

5.5.4 Test yield analysis

By using the Monte Carlo capabilities of design software and knowledge of test or measurement criteria, yield prediction analyses can be performed on designs that are described in appropriate models.

5.5.5 Maintainability analysis

Maintenance tasks that might be necessary, such as lubrication, cleaning, replenishment, calibration, failure diagnosis and repair, must all be analysed to ensure that they can be performed correctly by the people likely to be involved. Aspects that should be covered include physical accessibility, time to perform the tasks, skill levels, training requirements, special tools and equipment, and the need for special facilities.

5.6 SUMMARY

All of the methods described above are expensive in engineering time, and this can be a constraint on their effective application. They all involve the detailed consideration of many aspects, such as product characteristics in QFD, system simulation, LSA, failure modes in FMEA, SA and FTA, and process analysis. However, they can help to reduce overall project time and cost if they are applied effectively, particularly since the prime objective is always to identify and prevent or correct problems early. If product timescales are very tight, which is often the case, it is important that early decisions are made on which methods will be applied and how the results will be used. It might be appropriate to limit the scope of the analyses. For example, the QFD might be limited to a small number of critical design requirements, including variations if appropriate, and the LSA and FMEA to a few identified critical components rather than applied to all.

For most engineering projects a high-level QFD, design analysis using appropriate simulation software and FMEA should be the minimal approach. The top-down FMEA approach mentioned above can be particularly effective.

5.7   SOFTWARE FOR DESIGN ANALYSIS

Analysis of designs has traditionally been performed by specialists, requiring extra resources, time and money. However, the widespread availability of low-cost computing power has stimulated rapid evolution of engineering software, so that now engineers can, with appropriate training and management, perform most of the relevant analyses concurrently with other design tasks. The programs are becoming more powerful and easier to use, so that complex designs can be created and analysed thoroughly, effectively, accurately, quickly and at lower cost. The examples in this chapter show only relatively simple applications.

There is an increasing trend for programs to be made more comprehensive, so that the boundaries between, for example, software for drafting, mathematics, FEA/CFD, technology/physics, system simulation, virtual reality, etc., are disappearing, and there are considerable overlaps between the capabilities of many of the programs. Some specialise in particular fields, such as fatigue, circuit design, electromagnetics, etc., so the choice of which software is best for any application depends upon the technologies and problems involved.

An important result of the development of modern analysis software is the possibility of reducing or even eliminating the need for development testing. As discussed in earlier chapters, this is a very worthwhile objective, and projects are nearly always under pressure to reduce development costs and time. We will discuss this aspect in the following sections.

Suppliers of design analysis software, covering all of the methods described in this chapter, are listed on the book homepage (Preface page xvii).

5.8   LIMITATIONS OF DESIGN ANALYSES

All design analysis and simulation methods involve assumptions and simplifications that can, to varying degrees, generate erroneous or misleading results. The methods imply that we, and the software used to model the components and the system, know and fully understand all of the cause and effect relationships and interactions that affect performance, reliability, etc. Placing faith in the results of analysis can present risks which depend upon the design being analysed, the methods and software used, and the skill and knowledge of the engineers involved.

No design analysis software can deal with the whole range of possible operating stresses, environments, variations and degradation mechanisms that can cause failures. Inevitably the models are greatly simplified in relation to much of reality.

FEA software for mechanical stress analysis assumes (unless otherwise instructed) that material surfaces are smooth and not damaged, and that no other effects such as corrosion are present. Software for fatigue life prediction evaluates expected (average) lifetimes and variations around these, not the possible time to the first failure. The correctness of FEA outputs, whether applied to structures, CFD, electromagnetic effects, etc., depends on the correctness of the input descriptions such as surface conditions, the adequacy of the mesh being used, and understanding of the underlying mechanics and physics. Small errors or omissions in the mesh or other inputs can diverge and result in large errors in the outputs.

All component 'models' in EDA programs imply that the components are fully characterised, and the program user assumes that the models are all correct and complete. In particular, EDA software generally cannot simulate electromagnetic interference (EMI) effects on circuit performance, so it might not reveal that a circuit might fail because of layout, inadequate grounding, or under certain operating conditions.

Whilst it is always possible in theory to analyse the effects of variation by performing analyses with parameter values set at, say, tolerance limits or over tolerance ranges, and most CAE software includes facilities for tolerance analysis, it is often difficult and time-consuming to perform such analyses effectively. In particular, analysis implies that distribution parameters and interaction effects are known. As explained in Chapter 4, these aspects are often very uncertain.

The methods enable engineers to create and analyse complex and difficult designs quickly and economically. However, it is not safe to rely on the software to replace the need for knowledge and experience. Rather, for the most effective application of modern CAE methods engineers should possess thorough knowledge of the appropriate theories and of their practical application, and of the capabilities as well as the limitations of the analysis methods. Ultimately, only the actual hardware embodies the whole truth of the design, particularly aspects which might have been neglected or misrepresented in the analysis. *The need for testing is a direct consequence of the 'uncertainty gaps' that arise as a result of the limitations inherent in all design analysis.*

5.9 USING ANALYSIS RESULTS FOR TEST PLANNING

Since the analytical methods described indicate the aspects of designs that are critical or might present risks, they provide an excellent basis for optimising the test programme. Therefore the results of the analyses should be used to help to plan and prioritise the tests, and the engineers involved should be part of the test team. Test methods are described in the next four chapters, and we will discuss the management aspects of integrating the design and test activities in Chapter 14.

REFERENCES

1. Akao, Y., 1990, *Quality Function Deployment*, Productivity Press.
2. Hull, D. W., 1999, *Mastering Mechanics 1: Using MATLAB 5*, Prentice Hall.
3. Turcotte, L. H. and Wilson, H. B., 1998, *Computer Applications in Mechanics of Materials Using MATLAB*, Prentice Hall.
4. Attia, J. O., 1999, *Electronics and Circuit Aanlysis Using MATLAB*, CRC Press.
5. Bishop. R. H., 1997, *Modern Control Systems Analysis and Design Using MATLAB and SIMULINK*, Addison-Wesley.
6. Bishop, N. W. M. and Sherratt, F., 2000, *Finite Element Based Fatigue Calculations*, Appendix 3, NAFEMS.
7. Peyret, R., 1999, *Handbook of Computational Fluid Mechanics*, Academic Press.
8. Lonngren, K. E., 1997, *Electromagnetics with MATLAB*, International Science Publishing.
9. Mantooth, H. A. and Fiegenbaum, M., 1995, *Modeling with an Analog Hardware Description Language*, Kluwer Academic.
10. Banks, J. (ed.), 1998, *Handbook of Simulation*, John Wiley & Sons, Ltd.
11. O'Connor, P. D. T., 1995, *Practical Reliability Engineering* (3rd edn), John Wiley & Sons, Ltd.
12. Andrews, D. D. and Moss, T. R., 1993, *Reliability and Risk Assessment*, Longman (also published in USA by John Wiley & Sons Inc.).

6

Development Testing Principles

6.1 INTRODUCTION

Development testing is an integral part of the design process. Tests are performed to determine the extent to which initial designs achieve the specified performance and other requirements, to provide information on materials, components and processes that will be used, and finally to verify that the design of the product and of the processes by which it will be made and maintained is correct. Engineering design and test are nearly always iterative processes.

Chapter 1 introduced the basic categories of development testing, which are:

1. Functional performance testing, to demonstrate that the design (including design of the processes by which the product will be made) is capable of meeting the specified performance requirements. (This is also called '*design proving*' or '*proof of principle*' testing)

2. Reliability and durability testing

3. Contractual, safety and regulatory compliance.

In this chapter we will describe the first two categories. Safety and regulatory compliance testing will be described in Chapter 13.

A general difference between functional and compliance testing on one hand, and testing for reliability and durability (including the effects of variation), is that the former category needs to be performed only once, or maybe a small number of times or on a small number of items, whereas reliability/durability/variability testing usually involves multiple tests and test items, because of the wider uncertainty associated with these aspects. Another important difference is that we do not deliberately plan to cause failures in the first category, but we do in the second, as will be discussed in more detail later. For example, if we are setting out to demonstrate that a filter circuit design is capable of blocking a certain frequency range, then one test, carried out under expected operating conditions, would usually suffice. If, however, we want to know how production items will

97

perform when operated at conditions close to the environmental stress limits over long periods, the single design proving test will not help much.

Development testing evaluates the capability of the design of the product and of the processes which will be used to make it, and, when appropriate, to maintain it. The expression *test and evaluation* (T&E) is used to encompass the whole range of development testing in projects such as military systems: evaluation relates to testing of features such as operational effectiveness.

6.2 FUNCTIONAL TESTING

The requirements for functional test are as wide as the technologies and products that are encompassed by modern engineering. A few examples will illustrate the range.

- A new pressure transducer design: electrical output in relation to pressure input; temperature stability; linearity; sensitivity; maximum and minimum temperature operation; maximum pressure operation; response to over-pressure; operation under vibration; resistance to contamination.

- An electronic system employing digital, analogue and power technology: electrical performance, correct outputs over the range of inputs; operation over specified range of temperatures and power inputs; timing; EMI/EMC aspects.

- A new diesel engine: power, torque and speed relationships; fuel consumption/efficiency; low/high temperature starting; lubricant consumption and condition; vibration and noise output; exhaust gas composition.

- The Space Shuttle: launch performance such as engine thrust and control; in-orbit performance: control, crew environment, instruments, etc.; re-entry: control during atmospheric re-entry, re-entry thermal protection, etc.; landing: low-speed performance, undercarriage operation, braking.

Of course, for all of these designs an enormous amount of earlier testing will have been performed on materials, components and subsystems. The information gained will have been recorded in data sheets which will have been used by the designers, so, for example, the transducer and circuit designers might not necessarily have to test components such as output transistors, and the diesel engine and space shuttle designers might not test steels, aluminium alloys or transducers. We say 'might not', because it might also be necessary to test these items and materials as part of the test programme for the new design. This decision must be based upon the extent to which the information in the data sheets can be relied upon for the new application, and how critical is the function of the material or component. The designers must think about every item, and must decide whether and to what extent it should be tested. There can be no general rule, except that this thinking must be performed and the decisions made. The analysis methods described in the previous chapter should provide the basis for determining the appropriate tests.

The main testing technologies and methods are described in Chapter 7 (mechanical), Chapter 8 (electrical/electronic) and Chapter 9 (software).

6.3   TESTING FOR RELIABILITY AND DURABILITY: ACCELERATED TEST

In Chapters 2 and 3 we have reviewed how stresses, in the widest sense, can lead to failures, and how variations of strength, stress and other conditions can influence the likelihood of failure or duration (time, distance, cycles, etc.) to failure. In this section we will describe how tests should be designed and conducted to provide assurance that designs and products are reliable and durable in service. We will assume that the basic functionality has been achieved: that is, at least one unit has been demonstrated to work as specified and designed, and maybe also for as long as specified, as discussed above. What has not been assured, however, is that all units produced will work over the whole range of specified conditions and expected lifetimes, or at least that they will do so to a sufficiently high level of reliability. The first condition is basically deterministic: the item passes or fails the test, and we know when we have done enough testing.

However, testing for reliability and durability is fundamentally different. The reasons are that we never know what is the 'uncertainty gap' between the theoretical and real capabilities of the design and the products made to it, for the whole population, over their operating lives and environments. In Chapter 2 we discussed the uncertainties of stresses and strength, and in Chapter 4 we saw how these uncertainties are even further magnified by the effects of variations of environments and processes. The effects of these uncertainties can seldom be evaluated with confidence by any of the design analysis methods described in Chapter 5. How then can we plan a test programme that will reduce the uncertainty gap to an extent that we can be assured of reliability and durability, whilst taking due account of practical constraints like cost and time?

The conventional approach to this problem has been to treat reliability as a functional performance characteristic that can be measured, by testing items over a period of time whilst applying simulated or actual in-service conditions, and then calculating the reliability achieved on the test, as described in Chapter 4. For example, time of operation divided by number of failures is the estimated mean time between failures (MTBF). In the 1950s the US Department of Defense set up the Advisory Group on Reliability of Electronic Equipment (AGREE), which established test and data analysis methods for this purpose. However, these methods are fundamentally inadequate, for reasons that will be described in Chapter 12. The main reason is that they are based on measuring the reliability achieved during the application of simulated or actual stresses that are within the specified service environments, in the expectation (or hope) that the number of failures will be below the criterion for the test. This is the wrong answer to the problem expressed above.

The correct answer is straightforward: *we must test to cause failures, not test to demonstrate successful achievement.*

This concept is well accepted in a few test situations, particularly in mechanical strength testing. To derive the strength and fatigue properties of materials, samples are tested to failure. The test methods are described in Chapter 7. As explained in Chapter 2, we cannot accurately determine the strength of, say, an alloy or a plastic material by theoretical analysis, only by testing samples to failure. If we design a component using such a material, we can analyse the stresses using methods like FEA and we can calculate the strength using the material properties derived from the tests to failure. If the design is simple, and there is an adequate margin between stress and strength, we might decide that no further testing is necessary. If, however, constraints such as weight force us to design with smaller margins, and if the component's function is critical (like supporting an aircraft engine), we might well consider it prudent to test some quantity to failure. We would then expect that failures would occur only well beyond the expected maximum stress/minimum life, to provide an adequate margin of safety to take account of the known uncertainties and variations in this kind of design and application.

However, let us assume that an electronic system is being designed, and the specified maximum temperature for satisfactory operation is 40°C. At what temperature should the prototype be tested? Some inexperienced (and some experienced) engineers answer that 40°C should be the maximum test temperature, because any temperature above that would not be 'representative' of specified conditions. Therefore any failures that occur above that temperature would not be considered relevant.

However, suppose that a prototype was tested at 42°C and failed. Should we ignore this? Might this failure occur at 35°C on another unit built to the same drawings (effect of variability), or might it occur on this unit 6 months into the warranty period (effect of a time-dependent failure mechanism)? Might it occur at a combination of 35°C and a small, within-specification, increase in supply voltage? Can we really be sure that the failure at 42°C is not relevant, just because the thermal stress applied was not 'representative'?

If the failure occurred at a temperature 2°C above the specified limit, it is unlikely that it would be ignored (though this does happen). Suppose, however, that failure occurred at 50°C, or 60°C? At what stress do we decide that the level is so high that we can ignore failures? Should we even be testing at stresses so much higher than the maximum specified values?

The answer is that these are the wrong questions. The clue is in the earlier questions about the possible cause of the failure. When failures occur on test, we should ask whether they could occur in use. The question of relevance can be answered only by investigating the actual physical or chemical cause of failure. Then the questions must be asked:

1. Could this failure occur in use (on other items, after longer times, at other stresses, etc.)?

2. Could we prevent it from happening in use?

The stress(es) that were applied are relevant only in so far as *they were the tools to provide the evidence that an opportunity exists to improve the design. We have obtained information on how to reduce the uncertainty gap.* Whether the opportunity is taken is a management issue, in which other aspects such as cost, weight, time, etc. must be considered.

The uncertainty described above is shown in Figure 6.1. If we consider only one stress and the failures it might cause, the stress to failure distribution of production items might be as shown. As a simple example, this might be the operating temperature at which an electronic component malfunctions, or the pressure at which a seal begins to leak. As discussed in Chapter 4, the exact nature of this distribution is almost always uncertain, particularly in the tails, which are the most important areas as far as reliability and durability are concerned.

Suppose that the first test failure occurs at stress level L. At this stage we might have only a few items to test, maybe only one. We can state that the strength of this item represents a point on the distribution, but we cannot say whether it was an average strength item, a strong one or a weak one. The only way to find out

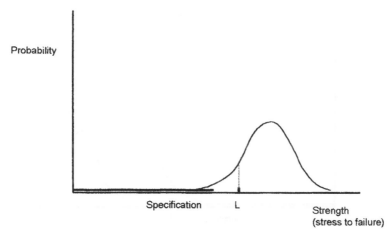

Figure 6.1   Stress, strength and test failures (1): strength vs. specification

the nature of the strength distribution is to test more items to failure, and to plot and analyse the results. Inevitably most of the items tested will be near to the average, because that is where most of the population will lie. Therefore, it is unlikely that any item tested will represent the weakest in the future population.

However, if we analyse the actual cause of the failure, by whatever means is appropriate, and take action to prevent its recurrence, then in effect we will move the strength distribution to the right. We still do not know its shape, but that is not what is important. We just want to move it out of the way. We are engineers, not theoretical scientists or statisticians, so *we can use high stresses in place of large samples.* Whilst scientific knowledge of the cause and effect relationships that affect reliability and durability is obviously necessary in order to create designs *and to determine how to improve them*, this is appropriate to determining where distributed values are centred, and sometimes the variation near the centre. In the electronics example, we might relocate the component to a part of the circuit which is cooler, or add a heat sink. For the seal we might change the material or the dimensions, or add a second seal, or reduce the pressure. The system will therefore be made more reliable.

For many items subjected to stress tests, particularly electronic systems, two types of failure can occur: transient (or *operating*) *failures and permanent failures.* If a transient failure occurs at some stress level the correct operation will be restored if the stress is reduced. Permanent failures are those from which the operation does not recover when the stress is reduced. For stresses like temperature, power voltage level, etc., which might have low or negative limiting values, failures might occur at high and at low levels, and these will of course have different physical causes from those at the high levels. For a population of items the stresses at which the failures occur will be distributed. The general case is illustrated in Figure 6.2.

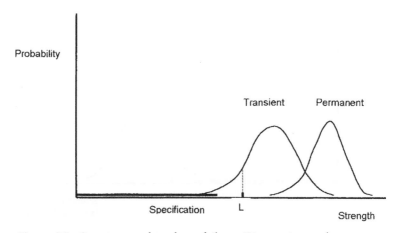

Figure 6.2 Stress, strength and test failures (2): transient and permanent

If the failure is due to a wearout mechanism, say wear of a bearing or fatigue of a component attachment, the horizontal axis of the distribution will represent time (or cycles), for any particular stress value. In addition to the uncertainty regarding the stress and strength, we now have the further uncertainty of time. Failure on test after time t at stress level L will represent one point on an unknown three-dimensional distribution (Figure 6.3). As Figure 6.3 illustrates, an important feature of wearout mechanisms is that the resulting distributions of times to failure become wider as damage accumulates, thus further increasing the uncertainty. To obtain a full understanding might require more testing, with larger samples. However, the same principle applies: we are not really interested in the shapes of distributions. We just want to make the design better, if it is cost-effective to do so.

In most engineering situations failures are caused by combinations of stresses and strength values, not just by one stress and one strength variable. Some might be time-dependent, others not. Using the examples above:

- The stresses applied to the electronic component could be a combination of high-temperature operation, high-temperature rate of change after switch-on, rate of on–off cycles, humidity when not operating, power supply voltage level and vibration. The resisting strengths might be mechanical integrity of the internal connections, thermal conductivity of the encapsulating material, absence of defects, etc.

- For the seal, the stresses and other variables might be oil temperature, pressure, pressure fluctuations, oil conditions (viscosity, cleanliness, etc.), shaft axial and radial movement, vibration, tolerances between moving parts, tolerances on seal grooves and seal dimensions, etc.

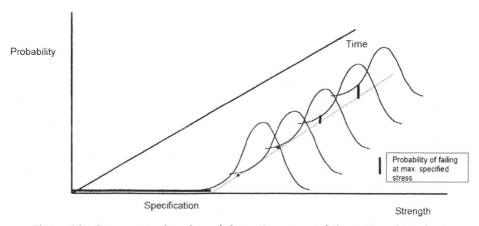

Figure 6.3   Stress, strength and test failures (3): wearout failures (time-dependent)

Between any two or more of these variables there might also be interactions, as described in Chapter 4.

Therefore, even for relatively simple and common failure situations like these, there is not just one distribution that is important, but a number of possible distributions and interactions. What might cause a transistor, capacitor or seal to fail in one application might have a negligible effect in another, or a component that has worked well in previous applications might cause problems in a new but similar one. This is how life is in engineering!

This reasoning leads to the main principle of development testing for reliability. *We should increase the stresses so that we cause failures to occur, then use the information to improve reliability and durability.* Clearly there will be practical limits to the stresses applied. These limits are set by:

- The fundamental limits of the technology. For example, there is no point in testing an electronic system at temperatures above the melting point of the solder used.

- The limits of the test capability, such as the maximum temperature of the test chamber.

The logic that justifies the use of very high 'unrepresentative' stresses is based upon four aspects of engineering reality:

1. The causes of failures that will occur in the future are often very uncertain.

2. The probabilities of and durations to failure are also highly uncertain.

3. Time spent on testing is expensive, so the more quickly we can reduce the uncertainty gap the better.

4. Finding causes of failure during development and preventing recurrence is far less expensive than finding new failure causes in use.

It cannot be emphasised too strongly that testing at 'representative' stresses, in the hope that failures will not occur, is very expensive in time and money and is mostly a waste of resources. It is unfortunate that nearly all standardised approaches to stress testing (these standards are discussed in Chapter 13) demand the use of typical or maximum specified stresses. This approach is widely applied in industry, and it is common to observe prototypes on long-duration tests with 'simulated' stresses applied. For example, engines are run on test beds for hundreds of hours, cars are run for thousands of miles around test tracks, and electronic systems are run for thousands of hours in environmental test chambers. Tests in which the prototype does not fail are considered to be 'successes'. However, despite the long durations and high costs involved, relatively few opportunities for improvement are identified, and failures occur in service that were not observed during testing.

An important point to realise in this context is that a failure that is generated by a stress level during test *might be generated by a different stress (or stresses) in service*. For example, a fatigue failure caused by a few minutes vibration on test might be caused by months or years of temperature cycling in service. The vibration stress applied on test might be totally unrepresentative of service conditions. Once again, though, the principle applies that the 'unrepresentative' test stress might stimulate a relevant failure. Furthermore, it will have done so much more rapidly than would have been the case if temperature cycling had been applied.

This approach to test is called *Highly Accelerated Life Testing* (HALT). HALT extends the principle of step-stress accelerated testing to the logical conclusions described above. The principle was developed by an American engineer, Dr Gregg Hobbs, and is fully described in Reference 1. In HALT we make no attempts to simulate the service environment, except possibly as the starting point for the step-stress application. No limits are set to the types and levels of stresses to be applied. We apply whatever stresses might cause failures to occur as soon as practicable, whilst the equipment is continually operated and monitored. We then analyse the failures as described above, and improve the design.

The HALT approach, and variations on it, are also referred to as (*highly*) *accelerated stress testing* (H)AST, *stress-induced failure environment* (STRIFE) *testing*, or *failure mode verification testing* (FMVT)® (Entela Corp.). Test facilities for these methods are described in Chapter 7.

Accelerated stress testing can provide quantitative cause and effect information when the mechanisms of failure are already understood (e.g. material fatigue), and when the tests are planned specifically to provide such information. We can perform statistically designed experiments (Chapter 4, and below), applying accelerated stresses to explore their effects. However, tests to provide such information require larger samples, more detailed planning and more time, and therefore would cost more, than would be the case in the accelerated test approach described above. We must decide whether we need the extra information that more 'scientific' or statistical tests can provide. In many cases the information from accelerated tests, as described above, coupled with engineering knowledge, is sufficient to enable us to take appropriate action to improve designs and processes. However, sometimes we need to obtain more detailed information, especially when the cause and effect relationships are uncertain.

6.3.1   Test approach for accelerated test

The approach that should be applied to any accelerated test programme for reliability/durability should be:

1. Try to determine, as far as practicable, what failures might occur in service. This should have been performed during design analysis and review, parti-

cularly during the failure modes, effects and criticality analysis (FMECA) and quality function deployment (QFD), as discussed earlier.

2. List the application and environmental stresses that might cause failures. Use Chapters 2 and 3 for guidance.

3. Plan how the stresses that might stimulate foreseeable and unforeseen failures can most effectively be applied in test. Set up the item (or items) to be tested in the test chamber or other facility so that it can be operated and monitored.

4. Apply a single stress, at or near to the design maximum, and increase the level stepwise until the first failure is detected.

5. Determine the cause and take action to strengthen the design so that it will survive higher stresses. This action might be a permanent improvement, or a temporary measure to enable testing to be continued.

6. Continue increasing the stress(es) to discover further failure causes (or the same cause at a higher stress), and take action as above. This approach is called *step-stress* accelerated testing.

7. Continue until all of the transient and permanent failure modes for the applied stress are discovered and, as far as technologically and economically practicable, designed out. Repeat for other single stresses.

8. Decide when to stop (fundamental technology limit, limit of stress that can be applied, cost or weight limit).

9. Repeat the process using combined stresses, as appropriate and within the equipment capabilities (temperature, vibration, power supply voltage, etc.).

The selection of stresses to be applied, singly or in combination, is based upon experience and on the hardware being tested, and not on specifications or standards.

The HALT approach can be applied to any kind of product or technology. For example:

- *Engines, pumps, power transmission units such as gearboxes, etc.:*

    — Start tests with old lubricants or other fluids (coolants, hydraulics, etc.), rather than new.
    — Run at low fluid levels.
    — Use fluids that are heated, cooled or contaminated.
    — Use old filters.

— Misalign shafts, bearings, etc.
— Apply out-of-balance to rotating components.

- *Electro-mechanical assemblies such as printers, document, material or component handlers, etc.:*

    — Apply high/low temperatures, vibration, humidity/damp, etc.
    — Use components with out-of-tolerance dimensions.
    — Misalign shafts, bearings, etc.
    — Use papers/documents/materials/components that exceed specifications (thickness, weight, friction, etc.).

- *Small components or assemblies such as electronic packages, mechanical latches, switches, transducers, etc.:*

    — Apply high/low temperatures, vibration, humidity/damp, etc.
    — Apply high frequency vibration by fixing to suitable transducers, such as loudspeaker coils, and driving with an audio amplifier.

When testing under such conditions the times to failure will nearly always be much less than expected under worst-case operating conditions. Therefore, no demonstration will be provided of expected in-service operating life. What we obtain instead is early knowledge of the features that limit reliability and durability, so that we can take action to improve the design. If a demonstration of durability is required this can be performed separately, after the HALT process. (Compare this with the case of the locomotive diesel engine 'type test' described on page 13.)

By applying stresses well in excess of those that will be seen in service, failures are caused to occur very much more quickly. Typically the times or cycles to failure in HALT will be several orders of magnitude less than would be observed in service. Failures which might occur after months or years in service are stimulated in minutes by HALT. Also, the very small sample usually available for development test will show up failure modes that might occur on only a very small proportion of manufactured items. *Therefore we obtain time compression of the test programme by orders of magnitude, and much increased effectiveness. This generates proportional reductions in test programme cost and in time to market, as well as greatly improved reliability and durability.*

As mentioned above, the basic idea of applying accelerated stresses is quite normal in mechanical stress engineering for critical components. In these situations fracture, possibly caused by fatigue, is usually the predominant failure mode being considered. For example, standard methods for pressure vessel testing (ASME Codes, British Standards) require that new designs must be tested at 2–4 times rated pressure. It is not expected that any units will fail. If failure occurs just below the test pressure the design will not be acceptable, but there is no requirement to continue to increase pressure to failure. However, for other technologies and at the systems level, there are usually many more potential causes of failure, and many of these are more uncertain and subject to greater variability in times to occur. This fact makes the step-stress test to failure approach even more relevant, not less so, for other technologies and for assemblies and systems.

Accelerated stress testing has been performed in some other technology and system applications for many years, but it has been unusual for the stresses to be extended considerably beyond the specified in-service values. Electronic systems have been subjected to accelerated temperature and vibration stresses, but typically to levels of the order of 20–50% above specifications, in development and for production units. This is called *environmental stress screening* (ESS). Other names have been used for the same approach, including STRIFE (stress + life). ESS methods have been standardised to some extent as a result of the guidelines published by the US Institute of Environmental Sciences and Technology (IEST) (Reference 2). These are discussed in Chapter 10. Also, apart from the fairly limited stress combinations applied in ESS or CERT, the stresses are usually applied singly. The objective is to simulate the expected worst-case service envi-ronments, or to accelerate them by only moderate amounts. The general principle usually applied to all of these methods has been to test the item to ensure that it does not fail during the test. *This is not consistent with the HALT approach.*

6.3.2 HALT and production testing

HALT does not only provide evidence on how to make designs more robust. It also provides the information necessary to optimise stress screens for manufacturing. The basic difference between the objectives of accelerated test in development and in manufacturing is that, whilst we try to cause all development test items to fail in order to learn how to improve the design, we must try to avoid damaging good manufactured items, whilst causing weak or defective ones to fail so that they can be corrected or segregated. The knowledge that we gain by applying the full HALT sequence, including the design ruggedisation, can be used to design a stress test regime that is optimised for the product, and that is far more effective than conventional production testing. This is called *Highly Accelerated Stress Screening* (HASS). *HASS provides the same benefits in manufacturing as HALT does in development, in greatly increasing the effectiveness of manufacturing screens whilst reducing test cost and time.* We will describe manufacturing testing in detail in Chapter 10.

Note that HALT and HASS represent an integrated approach to testing to ensure that both the design and the manufacturing processes will generate highly reliable products, at minimum cost and time. Conventional, separate, approaches to development and manufacturing tests do not ensure this integration, and therefore can result in much lower reliability and higher costs.

6.3.3   Common questions on HALT

Some questions that frequently arise when the HALT idea is described or first applied, and the answers, are as follows.

How many items should be subjected to HALT?

Answer: As many as can be made available, within project cost and other constraints. As discussed above, there can be no 'ideal' quantity from a statistical point of view. However, by testing more than one item we increase our chances of discovering design weaknesses, and we might also obtain more information on the important variables. HALT on one item is far more beneficial than not applying HALT, and the value of testing multiple items would usually begin to decrease after typically four or five. HALT should also be re-applied as appropriate to prove the effectiveness of design changes, and to ensure that new failure modes have not been introduced. It is also a good idea to apply HALT from time to time during production to ensure that no new weaknesses have arisen as a result of component supplier changes, process changes, etc.

How can reliability/durability values be demonstrated or measured using HALT?

Answer: They cannot. An accelerated stress test can provide such information only if the cause of failure is a single predominant mechanism such as fatigue, we know exactly what single type of stress was applied, and we have a credible mathematical relationship to link the two. Such relationships exist for some failure mechanisms, as described in Chapters 2 and 3. However, since HALT applies a range of simultaneous stresses, and since the stress profiles (particularly the vibration inputs) are complex and unrecorded, such relationships cannot be derived. In HALT we are trying to stimulate failures as quickly as possible, using highly 'unrepresentative' stresses, so it is impossible and misleading to relate the results to any quantitative reliability/durability requirement such as MTBF, MTTF, etc. However, if we take action on the failures, we can confidently assert that reliability/durability will be improved.

How can we be sure whether a failure generated by HALT might also occur in service?

Answer: By analysing the fundamental cause, thinking about it, and applying experience. There can be no 'scientific' proof that the failure will occur in service, when or how often. However, by study of the root cause (fatigue, overheat, chafing, etc.) we obtain information on how to prevent or delay its recurrence, and make the necessary changes. Sometimes experience exists of the failure having occurred previously in service, for example on earlier versions of the design, and this can be very powerful evidence both of the effectiveness of HALT in illuminating such failures in advance, and of the relevance of the failure during HALT to in-service reliability, and therefore the need to take action to improve the design. If doubt exists, *assume that the stimulated failure will occur in service.*

The product under test will not be subjected to vibration in service, so what is the point of applying high vibration stresses during HALT/HASS?

Answer: This is a frequently asked question, particularly in relation to electronic equipment which will be static in service, such as instrumentation, telecoms systems, etc. Again, the objective of HALT is not to simulate the operating environment, but to stimulate failures so that we are shown opportunities to improve the design. If the failures generated by vibration are the same as might be generated by, for example, temperature cycling or transport shocks, then they are relevant. In this type of situation, vibration often can show up design weaknesses much more quickly than will thermal cycling or shock application, since the rate at which energy is applied by vibration is so much higher.

Can the principles of HALT/HASS be applied to other conditions, environments, etc.?

Answer: Yes. We can use any methods that are realistic and practical to stimulate failures more quickly than they might occur in service. For example, we can test bearings, transmissions, hydraulic components, etc. using lubricants or fluids that are old or deliberately contaminated, we can deliberately remove protective devices, or we can test items with components that are deliberately outside tolerances or damaged. Note that all of these situations can occur in service, so we will obtain information on how the product withstands these kinds of misuse, in addition to information on how to make it more robust.

6.3.4 Overall benefits of HALT and HASS

- We generate robust designs and capable processes. Together these equate to *high reliability and durability.*

- *We reduce test time and cost, in development and in manufacture.*

- Since design and process weaknesses are demonstrated so quickly, nearly always within one day, corrective action can also be taken quickly. There is much less need for formal communication, since it is easy to arrange for the designers to be present when the tests are performed, and for them to participate in the testing. This reinforces teamwork and reduces costs and time.

- Designers are also given powerful lessons in how to avoid creating weaknesses in future, so that their future designs will be more reliable and durable even before the start of testing. Thus the capability of the design team is enhanced, resulting in even greater cost and time savings on successive product developments. *We generate continuous improvement ('kaizen') of design capability of products and processes.* This continuous improvement of design capability has a much greater potential for improving productivity and competitive position than continuous improvement of downstream manufacturing processes. *Kaizen* in manufacturing is an excellent philosophy, but *kaizen* of engineering design is even more effective.

6.4 TESTING FOR VARIATION: TAGUCHI METHOD

In Chapter 4 we explained the principles of statistical design of experiments (DoE) and the Taguchi method. In any design (product and processes) in which performance, reliability or durability might be affected by variations, they must be included in the test programme. If only one or a few variations are considered likely to have significant effects, the effects are understood, and there are no interactions between them, then conventional one-at-a-time tests, using the HALT approach, might be sufficient. However, if these assumptions cannot be safely made, then statistical experiments should be performed to determine which variations and interactions are significant.

The Taguchi method, starting with a brainstorm to identify the possible variations, their possible effects and possible interactions, as well as areas of uncertainty, is the best way to determine what tests should be performed and to optimise them in terms of cost, time and effectiveness. The brainstorm team must include all of the people whose work can influence the product and the processes, so designers, suppliers' engineers and production people must participate as appropriate. An essential further member of the team must be either an engineer who understands the statistical methods or a suitably qualified statistician who understands the engineering. Unfortunately there are not many people who combine these qualifications, so it might be necessary to find a suitable engineer or statistician and provide additional training.

It is sometimes stated that statistical experiments are not helpful in engineering development, on the grounds that engineers know all of the cause-and-effect relationships and can therefore analyse them using deterministic methods. For example, the effects of all of the tolerances, temperature coefficients, parameter drifts, etc. in an electronic circuit can, in principle, be analysed using circuit theory and CAE software. If such comprehensive knowledge does exist, and it often does, then there is indeed no justification for statistical experimentation. If, however, there is uncertainty about some variables or interactions, which is also often the case, then the Taguchi brainstorm should highlight these, and appropriate tests can be run and analysed. It is important to realise that the Taguchi method does not demand statistical experiments. It provides a framework for deciding whether such experiments might be helpful, and for optimising them in the engineering context. To put it another way: *the Taguchi method always works*.

Statistical experiments and the analysis of the results provide information on *how much* the variables and interactions affect parameters of interest. They do not provide explanations *why*. Only theoretical knowledge of the underlying science can provide such answers. However, the statistical information can be valuable in generating clues for better understanding and control, particularly of designs and processes that are not easy to optimise using non-statistical methods. Examples of these are high-speed/high frequency electronic circuits, complex precision mechanisms and solder processes in electronics manufacture.

6.4.1   DoE or HALT?

Statistical experiments and HALT are complementary approaches in development testing. Table 6.1 gives some guidance on which approach to select for particular situations.

Note that these are by no means clear-cut criteria, and there will often be shades of grey between them. We must decide on the most appropriate method or combination of methods in relation to all of the factors: risks, knowledge, costs, time.

Table 6.1   DoE/HALT selection

Important variables, effects, etc.	DoE/HALT?
Parameters: electrical, dimensions, etc.	DoE
Effects on measured performance parameters, yields	DoE
Stress: temperature, vibration, etc.	HALT
Effects on reliability/durability	HALT
Several uncertain variables	DoE
Not enough items available for DoE	HALT
Not enough time available for DoE	HALT

6.5   PROCESS TESTING

The processes of manufacture, and of maintenance when appropriate, must be tested to ensure that they are capable of being performed correctly and economically. We reviewed the process analysis methods in the previous chapter. The results of these should be used to plan the process tests.

Most process testing involves performance of the tasks, human or machine, and evaluating the results in terms of correctness, accuracy and timing. Where machines are involved the tests might include aspects such as their reliability, safety, durability, maintenance, etc. For items that will be produced in large quantities the manufacturing facilities, such as dedicated machines, assembly robots and test equipment, can represent a significant development programme, and the testing of these must be closely integrated with the development testing of the product to be manufactured.

All manual tasks (assembly, repair, etc.) should be performed under the expected conditions, with the people, tools and other facilities that will be used later. The people involved should be required to comment on the tasks, and their comments should be addressed in the same way as other test or failure reports. Where appropriate the tasks should be timed, and reviewed to determine whether they can be improved.

6.5.1 Process capability studies

For manufacturing processes that are variable, such as machining or electronic component fabrication, the (statistical) *process capability* (C_p) is defined as the allowable tolerance width divided by the $\pm 3\sigma$ range of the (variable) process. The value of σ is determined by running samples of the process and performing statistical analysis of the variation in the output. A C_p of 3 or more indicates a 'capable' process, which would create only a very low proportion outside the tolerance range. To take account of situations when the process mean value is offset from the centre of the tolerance range, the numerator is (tolerance range − offset), and the process capability is denoted as C_{pk}.

Statistical process capability analysis implies that the variation follows the normal distribution. As explained in Chapter 4 this is not usually a safe assumption, so the method should not be relied upon to provide mathematically exact results. However, the basic underlying concept that the process variation should be well inside the tolerance range is obviously correct.

6.6 'BETA' TESTING

The testing approaches described so far are all performed by the engineers working on the project. Therefore, it is often possible that aspects of the designs that are influenced by or are important to users might be overlooked. Examples are ease of use of a new photocopier, handling characteristics of a new vehicle, maintenance accessibility of an engine installation and use of a new computer program. An effective way of evaluating such aspects is to place pre-production models in service with users who are selected to represent typical customers. This is called *'beta'* (*β*) *testing*.

The beta test users must be carefully selected and instructed. Their responsibilities are to use the product as would typical future customers, and to comment on all relevant design aspects. The user obtains the advantage of free use of the product during the test, and this can be extended or discounts given on the purchase of the production version. The designers obtain early feedback on aspects that can influence customer preferences, and can easily be overlooked during testing in-house.

Beta testing is commonly used for products such as domestic and office equipment, cars, software, etc.

6.7 CONCLUSIONS

Testing is an integral and essential part of any engineering development programme. It is also difficult to plan and expensive to execute, and the outcomes in terms of influence on the product, timescale and cost are nearly always uncertain. If the design phase can be compared to the planning of a battle, then development testing is the battle. However, it is only a preliminary to the greater challenges that will be faced later, in terms of reputation, reliability, safety and costs. The development test programme must be seen as the final opportunity to add value to the product by investing wisely to ensure its success. It is essential, therefore, that the test plans and their execution follow sound basic principles of engineering and management. This chapter has discussed the engineering principles. We will discuss technology aspects in the next three chapters, and the management and integration aspects in more detail in Chapter 14.

REFERENCES

1. Hobbs, G., 2000, *Accelerated Reliability Engineering: HALT and HASS*, John Wiley & Sons, Ltd
2. IEST, *Environmental Stress Screening of Electronics Hardware (ESSEH) Guidelines*, Institute of Environmental Sciences and Technology.

7

Materials and Systems Testing

7.1 MATERIALS

In Chapter 2 we described the main ways in which materials can fail. In this section we will describe the methods used to test materials to determine their strength characteristics. These tests can be applied to samples of the materials used, in which case the data is applicable to any component made of the material. However, sometimes the tests are performed on finished components as well, particularly if the manufacturing processes can alter the material properties. For example, fasteners such as rivets and bolts are manufactured from materials with known strength, but tests must also be performed on finished items, since processes such as forging, machining and heat treatment can affect the final strength.

7.1.1 Strength

Strength test, either static or fatigue, can be performed on test machines which apply graduated static or cyclic stress to the specimen under test. Figure 7.1 shows a typical test arrangement for tensile testing of material specimens. Because of the wide variation in static and fatigue strength in nominally identical samples, a number of specimens are tested and the averages and spreads are measured. At this level, failure is nearly always easily apparent as the specimen fractures, buckles, etc.

For large components and structures, such as vehicle bodies, aircraft wings, and train bogies, special test rigs must be constructed, and the stresses are applied by hydraulic actuators. Figure 7.2 shows a rail passenger vehicle being tested. The stresses applied are controlled either by controlling the pressure applied, or by feedback using strain gauges or other sensors. For fatigue loading the cyclic load spectra are programmed into the control system. It is usually practicable to test only one or a few of such large items, so due allowance must be made for the uncertainty inherent in the results. Failure, or incipient failure, at the level of such assemblies is often manifested by crack formation and growth, so regular inspection is important.

Figure 7.1   Tensile test machine (courtesy Denison-Mayes)

Typical criteria for strength of critical designs, such as aircraft and automotive safety-related components, as described in relevant standards (see Appendix 2) stipulate a margin between the worst-case stress and the low 3σ value of the measured distribution of strength. It is important to bear in mind the extent to which this can be misleading, as discussed in Chapter 4.

Mechanical stresses and strengths can be predicted by analysis and with knowledge of the material properties. Modern computerised finite element analysis (FEA) methods have made this task easier, faster and more accurate, particularly for complex loading situations (Chapter 5). However, actual tests to failure must be performed if the consequences of failure are severe, since there can be uncertainties, errors and incorrect assumptions in the analyses.

7.1.2 Hardness

Hardness is measured by measuring the indentation in material specimens when a defined shape is forced on to the specimen with a defined load. Two commonly

Figure 7.2   Strength/fatigue test on rail passenger vehicle (courtesy TuV Product Service)

applied hardness tests are the *Brinell* test and the *Rockwell* test. The former uses a 10 mm sphere of hardened steel or tungsten carbide, and the *Brinell Hardness Number* (BHN) is calculated from measurement of the diameter of the indentation and the load applied. The Rockwell test uses a diamond cone or a steel sphere, depending on the hardness of the material being tested, and the depth of the indentation at a given load is measured. The *Vickers* and the *Knoop* hardness tests are similar. In each case the hardness number is an empirical value, with units of force/length^2.

7.1.3   Toughness

Toughness is measured by impact tests, in which the energy level that causes a prepared specimen to fracture indicates the material toughness. The *Charpy V-notch* test uses a specimen with a defined V-shaped notch machined across the centre. The specimen is placed in the jaws of the test machine, and it is impacted by a heavy pendulum. The initial and post-impact maximum heights and the pendulum mass enable the fracture energy to be calculated.

7.1.4 Wear resistance

Resistance to wear is measured by subjecting the materials to wear conditions (adhesive, fretting, abrasive, etc.) in suitable test rigs. The amount and nature of the wear are measured by examination of the surfaces and by weight comparisons.

7.1.5 Corrosion and chemical attack

Tests for the ability to withstand corrosion or chemical attack are performed by appropriate methods, such as salt spray, immersion, etc.

7.1.6 Data

For most materials and standard components the information on relevant properties is available from specifications, databooks, etc., so that it is not usually necessary to perform further tests of the kinds described above.

7.2 ASSEMBLIES AND SYSTEMS

Assemblies and systems must be tested for their ability to perform in the range of expected operating conditions, since they might be made using several different materials and processes, and there are often interfaces and interactions between materials and components that can create conditions not covered by the basic material and component data.

The main test stress stimuli and the methods of application are described below.

7.2.1 Temperature

Temperature testing is performed using thermal chambers. These can range in size from small bench-top units of less than 1 m^3 to very large facilities in which complete items such as cars, rail vehicles and military systems can be tested. High temperatures are generated using electrical heaters. Radiant heating may also be applied. Low temperatures are generated by refrigeration units, or by using liquid nitrogen.

Typical chambers provide temperature ranges between +60°C and −40°C. An important further performance characteristic is the rate of change of temperature that can be achieved, particularly for tests that require rapid or frequent temperature excursions. Powerful electrical heaters and liquid nitrogen cooling systems can generate rates of change of 20°C/min, and up to 100°C/min in special chambers such as HALT chambers. Another way of generating high rates of change, or *temperature shock*, is to use a two-chamber system, with one hot chamber and one cold one, and a rapid transfer of the item under test between the two chambers.

The important temperature and rate of change values are not the ambient values in the chamber, but those measured on the item under test. Depending upon the thermal inertias and conductivities within the test item there will be temperature gradients and lags, and these must be estimated and allowed for in planning the thermal tests. Temperatures can be measured using thermocouples, infra-red scanners, or thermally sensitive indicator tape.

7.2.2 Vibration and shock

Items can be subjected to vibration test by being secured to platforms that are vibrated. The vibration systems are variously called shakers, vibrators, exciters or thrusters, though vibrator is the ISO preferred term. The main vibration sources are as follows.

- Eccentric rotating weights. These are simple, low-cost systems.

- Electrodynamic (moving element with armature coil, suspended in a static magnetic field, and driven by alternating electric current). The frequency and intensity can be varied by the control system. Such systems can range in size and power from small vibrators with frequency ranges up to 10 kHz, to very large, water-cooled units with frequencies up to 2 kHz and thrust values up to 300 kN.

- Pneumatic pistons driven by a pneumatic servo system. *Failure mode verification test* (FMVT) (^{TM}Entela Corp.) shakers are single-axis systems using a pneumatic actuator with a much greater amplitude than electrodynamic vibrators (about 100 mm), making them more suitable for testing components and assemblies whose failure modes are more likely to be stimulated with lower frequencies and higher amplitudes (mechanical items, vehicle assemblies, etc.).

- Hydraulic actuator systems are used for heavy items such as vehicles.

- Multi-axis systems, as described later.

Figure 7.3 shows an electrodynamic vibrator, as used for evaluating squeak and rattle in automotive assemblies.

Vibration applied may be at a simple fixed frequency, over a swept range of frequencies, or simultaneously over a selected range of frequencies. *Random vibration* is vibration that is applied over a range of frequencies and amplitudes that vary randomly over time. (*Pseudo-random* vibration is what is actually

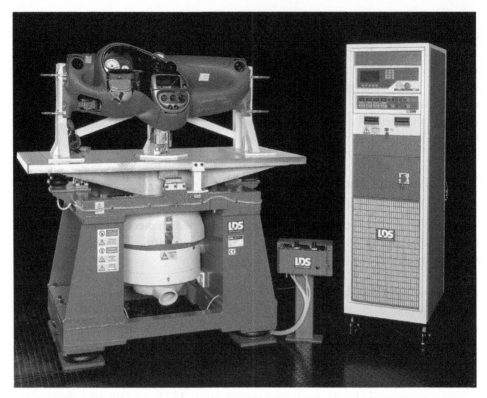

Figure 7.3   Electrodynamic vibrator system (courtesy Ling Dynamic Systems)

applied, as generation of true random vibration is not practicable, since it implies infinite displacement, etc., over an infinite frequency range.) By stimulating a range of vibrations simultaneously the existence of simultaneous resonances can be determined.

Power spectral density (PSD) is the term used to quantify vibration power at a frequency within the range. The unit of PSD is g_n^2/Hz (or, more correctly, g_n rms^2/Hz or the SI units m/s^2/Hz). The total force rating of a vibration system, over its frequency range, is expressed as g_n rms, kgf rms, N rms or lbf rms. It is the area under the curve of PSD vs frequency, such as in Figure 7.4.

Vibration control systems permit the vibration pattern and intensity to be planned and controlled. The vibration parameters measured on the unit under test (usually from suitably positioned accelerometers) are compared with the defined inputs, and the controller provides the required outputs to the drive amplifier of the vibrator. A typical random vibration test specification is shown in Figure 7.4.

Other vibration and shock inputs can also be applied, such as sine super-imposed on random, specific shock patterns, etc.

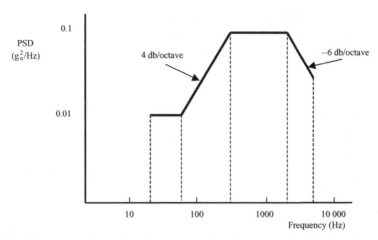

Figure 7.4 Typical random vibration test profile (Reproduced with permission of LDS Royston, UK)

For every item tested there will be frequencies at which the complete unit or parts of it will resonate, as described in Chapter 2. An important design objective is to ensure that any resonant frequencies will be beyond the range of expected operational input levels or low harmonics of these, so one of the objectives of vibration testing is to check for the existence of structural resonances within the operating environment. Swept sine inputs are applied, whilst the item is monitored using vibration transducers or a stroboscopic lamp.

Vibration inputs can be applied through different axes: a unit mounted vertically on a shaker will experience only vertical motion. For full exploration of vibration responses it is important to test the unit at different orientations. Three orthogonal axes are commonly tested, either separately by re-orienting the item, or with two or three axes simultaneously by having shakers connected to the different axes of the table.

Some vibration systems enable six degree of freedom (three axial directions and three rotational axes) random vibration to be applied simultaneously. These include:

- *Multi-axis shake tables* (MASTs). These are large systems that use multiple hydraulic or pneumatic actuators to drive the platform.

- Multi-axis repetitive shock systems, such as HALT/HASS vibrator tables, which are powered by a number of pneumatic hammers mounted on the underside. These input completely random, six degree of freedom (three axial, three rotational) vibration up to $0.1\ g_n\ \mathrm{rms^2/Hz}$ over a wide frequency range (10 Hz–10 kHz).

- The Cube (^{TM}TEAM Corp.), which is a freely suspended hollow rigid cube, with pneumatic actuators mounted onto the inside faces (upper and four sides). Items to be tested are fitted to the outside faces.

Fixtures and mounting

The design of fixtures for attaching items to the vibration platform is an important aspect of vibration testing, since the fixture must not alter the vibration inputs, e.g. by attenuation or by creating inappropriate resonances. The equipment must be rigidly mounted to the platform or fixture, so that it cannot bounce or become loose. Sensors such as accelerometers must be fitted to the appropriate parts of the item being tested, and care must be taken to ensure that they and their connecting cables are well secured.

Selecting vibration methods

Common questions on vibration testing are as follows.

- *Swept sine or random?*   Generally, swept sine vibration is the most effective way to investigate resonances, one at a time. However, random or pseudo-random vibration is the only way to investigate combined resonances, and is more effective for precipitating failures, either in development testing or in manufacturing screening. Single-frequency shake is cheap, but is very unlikely to be effective.

- *Single or multi-axis?*   If the item being tested will be constrained to only one direction of vibration in use, it might be practicable to apply single-axis vibration. Orthogonal axes can be vibrated by re-mounting the item, or by using a vibrator with a horizontal slip table. However, multi-axial simultaneous vibration is often a more realistic test, and is usually more effective in stimulating design and process-related failures.

- *Control and repeatability?*   On one hand, systems that generate controllable, repeatable vibration are considered preferable to those, such as pneumatic hammer shakers, that generate vibration controllable only in terms of total energy. On the other, it is argued that the most important result is product improvement, and that the precise conditions that show up design or process weaknesses are not as important. These aspects were discussed in Chapter 6. Both points of view are valid, depending upon the primary objectives of the tests. If it is important to determine the response of a design to specific vibration conditions, then a controllable system must be used. If the primary objective is to find weaknesses, in designs or in manufactured products, then the pneumatic hammer approach might be more effective, particularly when used as part of a combined environment accelerated test.

- *Vibration alone or combined environments?* See later.

Other aspects that should be considered are:

- The relevant characteristics of the design (mass, sub-assembly and component masses, attachments, etc., resonant frequencies, expected environments (transport, in-use), etc.)
- The kinds of failures that might occur, as described in earlier chapters, and how best to stimulate them
- The vibration test equipment that is available
- The overall economics (costs, benefits)
- Contractual or other project requirements.

Shock testing

Shocks are single events of high energy, or events that might result in short-duration high frequency vibration. Mechanical shock or impact can cause fracture; repeated shocks can cause fatigue, leading to fracture. Shock tests are performed by dropping the item onto a rigid base, by using shock test machines which allow defined accelerations to be applied, or by using more dramatic methods such as crash tests on vehicles.

The VIBKIT98™ (LDS) software is an effective tool for determining vibration and shock test parameters.

Reference 1 covers the subject of vibration and shock in detail.

7.2.3 Other conditions

There is a wide range of other test conditions that must be considered and applied, as appropriate to the product and its operating environment. These include the following.

- *Aerodynamic.* Items such as aircraft and missiles must be tested to determine properties such as aerodynamic stability, manoeuvrability and drag. Other vehicles are tested to determine drag, and structures are tested to determine stability and wind loads. Wind tunnels are used for aerodynamic testing, but computational fluid dynamics (CFD) software (Chapter 5) is being used increasingly to supplement or even replace the need for wind tunnel testing.

- *Acoustic noise.* Acoustic noise can affect the performance and reliability of systems that are exposed to very high noise levels, such as from jet engines and rocket exhausts. Noise can induce damaging vibrations on circuit cards, instrumentation, body panels, etc. Noise can also interfere with the performance of human operators, and can affect the comfort of people such as vehicle passengers. Therefore tests can be conducted that submit items to generated noise, using sound generators, as well as measure the noise generated by the item under test. In the automobile industry the term noise, vibration and harshness (NVH) is used to embrace the users' perception of vehicle performance, as affected by soundproofing, engine and powertrain balance, smoothness of operation of components such as door locks, gear changes, etc.

- *Acceleration.* Some systems and components must operate under high acceleration, in applications such as military aircraft, missiles and spacecraft. Weightless (zero-g) operation is also characteristic of such applications. High values of continuous acceleration can be generated using centrifuges. Zero-g tests can be performed only by dropping or by flying in zero-g trajectories, or by placing items in earth orbit.

- *Solar heating.* Systems that must withstand heating from solar radiation (outdoor equipment, spacecraft, etc.) are tested in temperature chambers that include high-intensity lamps.

- *Low pressure, space vacuum.* Aircraft, missiles and spacecraft must operate in low atmospheric pressures or in vacuum conditions. Low pressures can lead to arcing in electrical systems, as the dielectric strength of the air insulation is reduced, which is why aircraft and spacecraft electrical power systems operate at relatively low voltage levels (28 V DC, 115 V AC). Therefore low pressure or vacuum tests must be performed on such systems, using vacuum chambers.

- *Electromagnetic radiation.* Electromagnetic radiation aspects of testing are described in Chapter 8.

- *Lightning effects.* The transient electric currents and electromagnetic effects that can be induced by lightning strikes are simulated in special high-voltage lightning simulators.

- *People (and other animals).* The ways that people, including users, operators, maintainers and any others, interact with the system being developed must be included in the test programme. This aspect includes ergonomics (ease of viewing, understanding, operating, maintaining, etc.), forces applied to items like switches and controls, possibilities of accidents such as spilling coffee, dropping, incorrect operation, etc. Other animals are sometimes also part of the operating environment, for example rodents that eat cable insulation or enjoy sheltered accommodation in machinery, etc.

- *Other environments.* Other environmental stresses that can be applied using suitable chambers or other facilities include:
  — Humidity, condensation and freezing
  — Rain and snow
  — Mould growth
  — Sand and dust
  — UV and other radiation.

Stresses can be applied separately, using specialised facilities for each. As far as is practicable and cost-effective, operating and environmental stress conditions should be applied in combined ways that will stimulate the kinds of problems and failures that could occur in actual service. The combined effects of stresses are often more damaging, due to interaction effects, than separate application, and are often more representative of actual service conditions.

Of course, all of these conditions can also be applied by submitting the equipment to actual weather and other service conditions. For example, it is normal practice for new car prototypes to be driven in arctic and desert conditions. However, environmental chambers enable severe conditions to be applied in controlled ways, and usually much more economically. They cannot always reproduce the full range of actual conditions, particularly combined conditions. Also, there are some systems, such as engines and complete vehicles, that cannot be fully operated in typical environmental chambers. On the other hand, stress conditions that are more severe than would otherwise be experienced during development or in use can usually be applied. Therefore the test programme must be planned to make the optimum use of chamber tests and external tests.

Some of the environmental test facilities described above can be used to apply accelerated stresses, if they have the necessary performance. However, most of them can apply stresses (primarily temperature rate of change and vibration) that are lower than the stresses needed to implement the HALT approach effectively, and most enviromental chambers have limited capabilities to apply the kinds of multiple stresses that make the HALT principle truly effective. The reason for this is that they were developed in response to the conventional requirements to simulate operating environments, rather than to stimulate failures.

7.2.4 Combined environments

Single-environment testing is appropriate for evaluating specific responses. As discussed in Chapter 6, combined environmental stresses are usually more effective than single-environment tests, both for forcing design improvements and for manufacturing screening. The most common equipments for applying

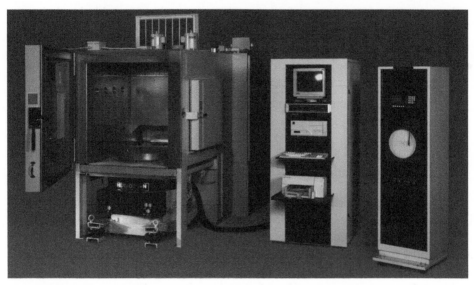

Figure 7.5   Combined environment test facility (courtesy Thermotron Industries)

combined environmental stresses to units under test are thermal test chambers which include other capabilities. These can be used to apply thermal and other stresses, as well as electrical power and monitoring to the unit under test (UUT). There are many types and configurations of such chambers. Using combined environments, particularly temperature cycling and vibration, to assess reliability using these types of system is sometimes called *combined environment reliability testing* (CERT). Another common name is *AGREE testing*, after the Advisory Group on Reliability of Electronic Equipment, set up by the US Department of Defense in the 1950s to develop methods for reliability testing for military electronic systems. Figure 7.5 shows an example. Some chambers include further capabilities, such as low pressure/vacuum, humidity control, etc.

CERT chambers generally are capable of applying stresses that simulate expected environmental conditions, rather than accelerating them.

7.2.5   Facilities for accelerated test

The only effective way to apply accelerated stress tests (Chapter 6), in most cases, is to use test facilities that have been designed for the purpose. Accelerated vibration test systems were described earlier. Facilities for accelerated combined stress applications are also available, and Figure 7.6 shows an example. The main differences between a conventional CERT chamber and a HALT (or similar principle) facility are as follows.

- *Temperature rate of change:* CERT chambers typically provide up to 20°C/min, whereas the HALT chambers can generate up to 100°C/min.

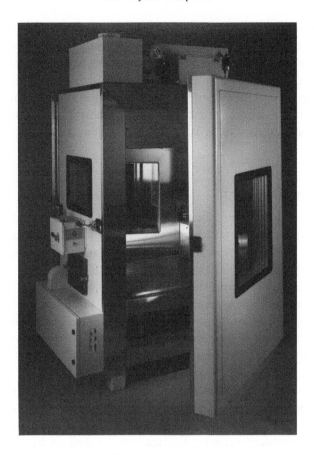

Figure 7.6   HALT/HASS test chamber (courtesy Qualmark)

- *Vibration:*   CERT vibration systems typically supply single-axis vibration up to 0.001 g_n rms^2/Hz. The HALT chamber applies random, six degree of freedom (three axial, three rotational) vibration up to 0.1 g_n rms^2/Hz over a wide frequency range (10 Hz to 10 kHz).

However, it is important to appreciate that there are overlaps between the performance capabilities of the different test systems, in relation to the stresses that could be accelerated, and there may be alternative ways of applying stresses. This was discussed in Chapter 6.

7.3   SYSTEM ASPECTS

7.3.1   Power and rotating machinery

Engines, electric motors, gearboxes and rotating power transmission systems are tested on test stands designed to accommodate the equipment, and to connect them to dynamometers or other power-absorbing systems. The main characteristics that are usually measured in machinery and rotating system tests, as appropriate, are power output, efficiency, vibration/noise, and durability.

Dynamometers can be frictional or fluid brakes, or electric generators that supply the output either to resistor load banks or into the electric utility power system. For vehicle power train testing the dynamometer is driven by the vehicle wheels, and simulated road or rail conditions (friction, roughness) are sometimes also applied.

7.3.2 Fluid systems

Hydraulic and pneumatic components and systems must be tested to ensure that they are safe and reliable under static and operating conditions. Since they operate at high pressures (typically 200–400 bar for hydraulics, 3–20 bar for pneumatics), safe operation demands that the margin between operating and rupture pressures is large, and these are specified in standards as described in Chapter 13. Therefore these requirements dictate the maximum test pressures that must be applied. In addition, tests are necessary to check for performance, leaks and deterioration. Components and systems can be tested on specially constructed test rigs, in environmental chambers, and as part of the overall system (vehicle, aircraft, power tool, etc.).

The main performance characteristics measured in fluid system and component tests, as appropriate, are rupture and leak pressures, flow rates, absence of induced vibration, pressure pulses or cavitation (pressure accumulators can be used to absorb transient pressure fluctuations, as well as to provide constant pressure during operation of actuators), reliability and durability.

7.3.3 Humans

Humans are part of many systems, for example as drivers, pilots, passengers, operators and maintainers. Sometimes it is necessary to test the human components of systems, or the 'man–machine interface' (MMI). Recent well-known examples of human (and sometimes other animal) testing are:

- The ability of man to withstand high 'g' forces, zero-g environments and reduced atmospheric pressure in aerospace applications.

- The survivability of drivers and passengers in accidents, and the effectiveness of protection systems such as energy-absorbing structures, seat belts and airbags. Instrumented anthropomorphic dummies are used for this work.

- Ability to cope with high work rates, stress, fatigue, boredom, etc. (pilots, racing drivers, process operators).

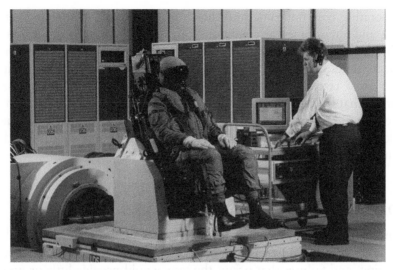

Figure 7.7 Military aircraft ejection seat and occupant being subjected to vibration test (courtesy TuV Product Service)

Subjects may be placed in appropriate simulated or actual environmental situations such as decompression chambers, simulators, vehicles, etc. (Figure 7.7).

We do not (at least not yet!) test humans in their own right as a development or production activity. However, testing for maintenance (diagnosis, recovery, etc.) is the job of doctors, and the equipment they use is often complex. Correct and safe functioning of heart pacemakers, dialysis machines, breathing ventilators, scanners and other medical systems requires that humans are included in the development tests.

Test instrumentation for human/animal testing includes:

- Electrocardiographs for heart function
- Electroencephalographs for nervous system monitoring (brain and nerve impulses, eye muscle movement, etc.)
- Chemical analysis (spectral analysis, etc.) of body fluids
- X-ray, ultrasonic and NMR imaging.

7.3.4 Large systems

Large systems, such as trains, aircraft, agricultural machines, some spacecraft, military systems, factory automation, process plant, etc. cannot be tested fully in fixed installations such as test chambers. System-level tests can be performed only in the operating environments, and therefore the scope for applying different operating and environmental conditions can be restricted. System-level testing for such systems is also very expensive. Where appropriate, subsystems and components should be tested as described earlier (see also Chapters 1 and 14), and the system tests must then be planned to cover the conditions that are important at that level. For example, a motor might be tested and shown to work at low temperatures, but only a system-level test might show that it overheats if snow blocks the air cooling intake. Reference 2 describes testing methods and management for systems.

7.4   DATA COLLECTION AND ANALYSIS

Collecting and analysing the data from testing is obviously an essential task: without data on test conditions, the behaviour of the items being tested, their configuration, and failures that occur, the tests would provide little or no useful information.

For relatively simple tests and simple items being tested, data can be collected manually. However, for more complex systems and tests data collection can be automated. *Data acquisition (DAQ)* systems and software are used when testing systems such as engines, vehicles, electronic systems, etc., when a large amount of data is generated in a short time. The data can be stored in computer memory or in media such as magnetic tape. Analysis can be performed in real time or later, depending upon the type of test and the software used.

Data collection and analysis methods are described in Chapter 12.

7.5   STANDARD TEST METHODS

The best known standardised methods for testing are described in Appendix 2. However, it should be noted that most of the standard tests relate to simulated environmental conditions, rather than accelerated stresses, and most tests are of a single applied stress. Therefore, they often do not represent realistic service conditions, nor do they always provide all the information that might be needed to ensure reliability, durability and safety.

7.6   TEST CENTRES

Large companies, such as automotive, aerospace and other systems manufacturers, usually operate test centres at which a wide range of test facilities are available. Government and independent test centres also exist. In addition to the kinds of facilities described above, those described below are available, depending on the main product types to be tested:

- Test tracks for vehicle test running. These usually have sections that simulate different road and off-road conditions, such as normal highway, rough pave, potholes, cross-country, etc.

- Extreme climatic conditions, such as desert heat, tropical damp or arctic cold.

- Instrumented ranges and related facilities for testing aircraft and weapon systems.

- Space platforms (satellites, Space Shuttle) for long-term testing in extra-atmospheric/zero-g conditions.

7.7 INFORMATION

Information on suppliers of the equipment and services described in this chapter is available on the book homepage (Preface page xvii).

REFERENCES

1. Harris, C. E. and Crede, C. E., 1976, *Shock and Vibration Handbook*, McGraw-Hill.
2. Reynolds, M. T., 1996, *Test and Evaluation of Complex Systems*, John Wiley & Sons, Ltd.

8

Testing Electronics

8.1 INTRODUCTION

Electronic components and systems must be tested during development, for the
fundamental reasons described in Chapter 1. In this chapter we will describe
circuit test technologies and equipment, design of circuits to optimise testability,
testing of components, and testing for EMI/EMC.

Electronic circuits can be either analogue or digital. Analogue circuits include all
those that have inputs and outputs that are fixed or can vary over a range, such as
radio and radar transmitters and receivers, amplifiers, filters, power supplies, etc.
Digital circuits are those that have digital inputs and outputs, i.e. logical 0 and 1
values. Nowadays there are many circuits that combine both types of function, but
for test planning the analogue/digital classification is a convenient starting point.

Initial development testing of electronic system designs is often performed on
manually assembled circuits that are not fully representative of the production
versions. These are known as "breadboard" circuits. Breadboard tests provide
only very limited information in relation to the effects of variation and environmental conditions, including EMI. However, they are useful (in conjunction with
simulation) for early design proving.

Test methods and economics appropriate to production and in-service support
will be described in Chapters 10 and 11.

8.2 CIRCUIT TEST PRINCIPLES

8.2.1 Analogue

Testing analogue circuit functions and parameters involves a range of measurement technologies and instruments. Some of the most important and frequently
applied tests are:

- Current, voltage potential, and resistance

- Gain, impedance, waveforms and other aspects of AC circuits

- Signal characteristics, such as frequency, gain, power, distortion, jitter, etc., for
  audio frequency (10 Hz to 10 kHz), radio frequency (RF) (10 kHz to 1000 MHz)
  and microwave (1000 kHz to 100 GHz) and high-speed digital circuits.

8.2.2   Digital

The basic approach to testing digital circuits is to check that the output logic is correct for each possible input state. As a simple example, consider a two-input AND gate (Figure 8.1). The possible input conditions and the correct outputs for each of these is shown in the *truth table*. In principle, then, it is simple to test such a device for all possible input conditions and faults[1]. The fault conditions in this basic approach are inputs or outputs stuck at 1 (SA1) and stuck at 0 (SA0). So long as none of the internal gates are SA1 or SA0 the device will function correctly, from the logic point of view.

In this case four test conditions, or *vectors*, are sufficient to test all operating conditions. For any logic network, the number of vectors necessary to show correct operation is 2^n, where n is the total number of possible input conditions. For a simple digital circuit this presents no problem, but as device complexity increases, the time taken to perform such a test can become too long to be performed practically and economically in mass production.

Not all devices are based upon simple *combinational logic*, in which the outputs follow the input conditions. In devices such as memories and processors, or complex circuits that include such functions, the outputs depend upon the inputs as well as on data flow and memory locations of stored data. This is *sequential logic*. Alternative test methods have been developed for these devices, such as pattern sensitivity tests for memories.

There are other practical limitations of the basic SA fault model approach to testing. There are other kinds of fault condition that can occur, such as a stuck at input condition, in which an output follows an input logic state. Fault occurrence can also be dependent upon other factors, such as timing, speed, interference, temperature, or voltage supply, as described in Chapter 3. However, the SA1/SA0 model remains the basis for most digital circuit functional testing.

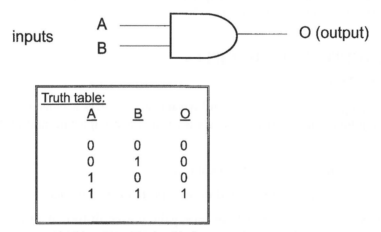

Figure 8.1   Truth table for two-input AND gate

1. The term "fault" is generally used in electronics testing to cover defects and failures.

Dynamic logic is that which requires a refresh or 'keep alive' clock. The term '*dynamic testing*' is often used to refer to 'at speed' testing. All digital circuits have speeds (clock rates) above which they will not operate correctly.

Reference 1 is a good introduction to digital circuit testing.

8.3 TEST EQUIPMENT

Circuits can be tested with manual test equipment or with *automatic test equipment* (ATE). Manual test equipment is used primarily during development, since at that stage it is not important to minimise test time, and greater flexibility is possible.

8.3.1 Manual test equipment

The main types of manual test equipment are as follows.

- Basic instruments, such as 'AVO' meters (amps, volts, ohms) or digital multimeters (DMMs), power meters and transistor testers are commonly available.

- Oscilloscopes, spectrum analysers and waveform generators measure, compare and analyse voltage and current levels and waveforms, and provide inputs so that circuit responses to input waveforms can be analysed. Most oscilloscopes enable two channels to be displayed.

- Logic analysers are similar to oscilloscopes but specialised for displaying digital pulse streams. They enable tests of digital inputs and outputs and timing to be performed.

- Special instruments, such as radio frequency testers, distortion meters, high voltage testers, optical signal testers, etc.

Figure 8.2 shows some typical modern manual test instruments.

Computer-based testing uses software that enables PCs to emulate test equipment. The PC is connected to the circuit to be tested via a databus and a data acquisition adaptor card. The PC screen becomes the instrument display, and the tests are controlled from the keyboard and mouse (Figure 8.3). Computer-based testing can reduce the costs of test equipment , since one PC can perform a range of test functions, and test results can be easily stored, analysed and transmitted. The microprocessors in PCs also enable tests to be performed faster.

8.3.2 Automatic test equipment

Automatic test equipment (ATE) is used for testing manufactured circuits, and also for in-service fault-finding and for testing repaired units. The main types of ATE for assembly testing are discussed in the following paragraphs.

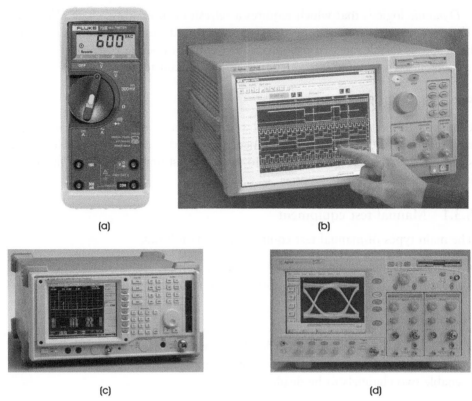

(a)

(b)

(c)

(d)

Figure 8.2 Test instruments: (a) DMM (Fluke); (b) Logic analyser (Agilent); (c) RF tester (IFR); (d) Optical signal tester (Agilent)

Vision systems

Vision systems refer generically to inspection systems that acquire an image and then analyse it. They do not actually test circuits, but they have become part of many production test sequences because of the great difficulty of human inspection of the large numbers of components, solder connections and tracks on modern circuits. *Automatic optical inspection* (AOI) machines are capable of scanning a manufactured circuit and identifying anomalies such as damaged, misplaced or missing components, faulty solder joints, solder spills across conductors, etc. One new system enables magnified images of solder connections under BGA packages to be viewed and analysed (Figure 8.4). X-ray systems (AXI) are also used, to enable inspection of otherwise hidden aspects such as BGA solder attachments and internal component problems. Other technologies, such as infra-red and laser scanning, are also used. Modern vision systems include software that enables criteria to be set for acceptability of features such as solder joints, part alignment, etc. and to store, update and compare images.

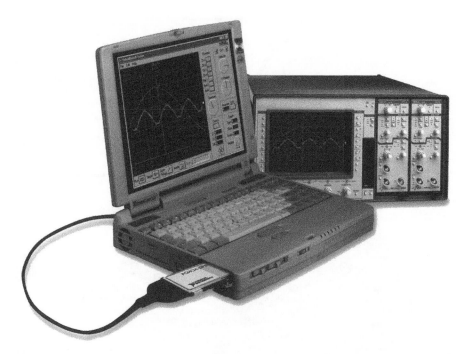

Figure 8.3 PC-based oscilloscope with conventional instrument (courtesy National Instruments)

In-circuit testers (ICT), manufacturing defects analysers (MDA)

ICT tests the functions of components within circuits, on loaded circuit boards. It does not test the circuit function. The ICT machine accesses the components, one at a time, via a test fixture (sometimes referred to as a 'bed of nails' fixture), which consists of a large number of spring-loaded contact pins, spaced to make contact with the appropriate test points for each component, for example the two ends of a resistor or the pin connections for an IC (Figure 8.5). The unit under test (UUT) is forced on to the fixture, and the ATE software then drives tests of all the components, such as resistance values, digital truth tables, gains, continuity of tracks, etc. The tests are all applied at low voltage levels, so that no components are damaged. ICT does not test circuit-level aspects such as tolerance mismatch, timing, interference, etc. Also, the test applied to ICs is only a basic test of relatively few logic functions. Therefore if a circuit passes ICT it is not ensured that it will function correctly. ICT can also provide effective diagnostic back-up for units that fail functional test, as described next. MDAs are similar but lower-cost machines, with capabilities to detect only manufacturing-induced defects such as opens, shorts and missing components: justification for their use instead of ICT is the fact that, in most modern electronics assembly, such defects are relatively more common than faulty components. MDAs do not apply power to the circuit under test, but checks for impedance matching in

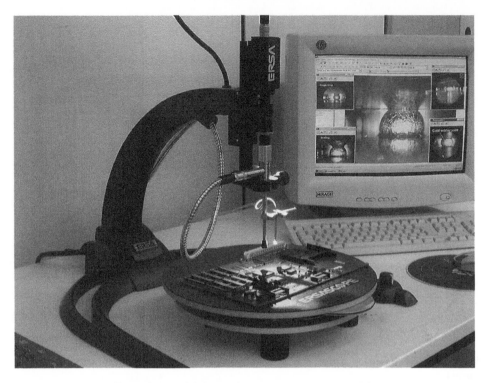

Figure 8.4   BGA inspection system (courtesy ERSA)

comparison with a known good circuit. ICT starts by applying power-off tests, then applies power for the component tests.

The majority of ICTs and all so-called 'high end' ICTs include some functional testing capability (mostly optional) that can perform common measurements such as voltage, current, time, frequency, etc. Special-purpose instruments can also be interfaced externally or internally in the form of VXI and PXI instruments on a board. In other words there can be a grey area between ICT and functional test. Also, for the simpler digital devices, where a full truth table test can be performed, the ICT performs a functional test. Fixtures can be built with dual-level probes such that, following a full nodal ICT test, the ICT probes can be removed (to prevent any loading of the circuitry), leaving a smaller number of functional test probes in contact with the board.

Most ICTs can also include a boundary scan capability, as described later.

Flying probe testers

Flying probe testers (also called *fixtureless testers*) access test points on the circuit using probes that are rapidly moved between points, driven by a programmed high-speed positioning system. The advantage over ICT is the fact that the probe programming is much less expensive and more adaptable to circuit changes than

Figure 8.5   In-circuit tester (ICT) (courtesy Agilent)

are expensive multi-pin ICT MDA adaptors, which must be designed and built for each circuit to be tested. A consequent advantage is that they can be used for testing during development as well as in production. They can also gain access to test points that are difficult to access via a bed-of-nails adaptor. The disadvantage is increased test time; however the latest flying probe testers are closing this gap (Figure 8.6).

Functional testers (FT)

Functional testers access the circuit, at the circuit board or assembly level, via the input and output connectors or via a smaller number of spring-loaded probes. They perform a functional test, as driven by the test software. Functional testers usually include facilities for diagnosing the location of causes of incorrect function. There is a wide range, from low-cost bench-top ATE for use in development labs, in relatively low complexity/low production rate manufacture, in-service tests and in repair shops, to very large, high-speed, high-capability systems. The modern trend is for production ATE to be specialised, and focused at defined technology areas, such as computing, signal processing, etc. Some ATE for circuit testing during manufacture includes combined ICT and FT, and then the ICT functions are performed first, followed by FT. Figure 8.7 shows a typical modern test station.

Figure 8.6 Flying probe (fixtureless) tester (courtesy Scorpion Technology)

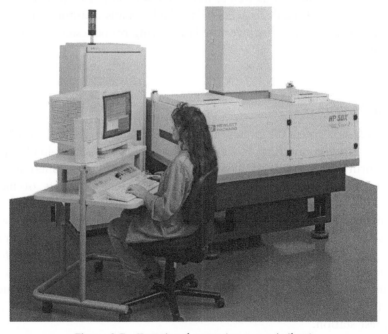

Figure 8.7 Functional tester (courtesy Agilent)

'Hot rigs'

A *'hot rig'* is a system which is used as a test facility for the sub-assemblies from which it is constructed. For example, a communications switch system consists of several subsystems, mostly on printed circuit cards, mounted in a cabinet. Any of the subsystems can be tested by being inserted into a cabinet which contains the other subsystems, and then testing the system. 'Hot rigs' are often used for testing in the manufacture of systems, particularly when the quantities produced are relatively low so that test time is not critical. They are also used in repair depots, for testing repaired items.

Special testers

There is also a wide range of specialised test equipment, such as power supply testers, cable testers, communication system testers, etc.

ATE is a very fast-moving technology, driven by the advances in circuit performance, packaging and connection technology and production economics. Reference 2 provides a good introduction, and Reference 3 describes ICT.

Manufacturing test options and economics are described in Chapter 10.

8.4   TEST DATA ACQUISITION

Modern electronic test systems use standard databus systems for test data acquisition (*DAQ*). These convert analogue signals to digital and control the data transfer between the UUT and the tester. IEEE488 was the first test databus standard, and most modern instruments include it. It is also called the *General Purpose Interface Bus* (GPIB). The *PC Interface* (PCI) bus or *PCI Extensions for Instruments* (PXI) bus are versions used for PC-based testing.

Larger functional test ATE systems use the VXI bus standard, which allows higher test speeds and greater accuracy.

Stand-alone *dataloggers* can be used to collect test data from equipment being tested, for later downloading to PCs for review and analysis. Dataloggers are useful at remote sites, in vehicles, and in other situations where it is not practicable or economic to set up more bulky instrumentation.

8.5   DESIGN FOR TEST

8.5.1   Test capability

Because production testing of complex circuits is so expensive and difficult, it is essential that they are designed to enable the ATE to perform effective testing. This is called *design for test* (DFT). The ATE must be able to:

- Confirm correct operation of good units
- Not classify good units as faulty
- Detect faulty units
- Diagnose the causes of faults.

The probabilities of these functions being performed correctly are called *test coverage*. There will nearly always be some types of fault which the ATE will not be able to detect and/or diagnose, particularly if test time is limited, so coverage will never be 100 percent. However, careful DFT can ensure coverage in the region of 95–98 percent for typical modern circuits, compared with 50–80 percent for poorly designed ones.

DFT must take account of the types of ATE that will be used, their performance (function, speed) and constraints, as well as quantities of the circuits to be manufactured, criticality of the circuit function, test costs and the downstream costs of undetected faults. DFT must also take account of the faults to be targeted. (The main causes of electronic component and system failure were described in Chapters 2, 3 and 4.) These aspects will be discussed further in Chapter 10.

To test an electronic circuit, it is necessary that the ATE is able to perform the following functions:

- *Initialise.* The ATE must put the circuit into a known initial state, for example digital inputs and outputs at known states, analogue levels at known values, etc.

- *Control.* Inputs must be generated, to stimulate changes on the outputs.

- *Observe.* The ATE must be able to detect the changes in output states and values caused by the presence of design errors (tolerances, etc.) or manufacturing faults.

- *Partition.* It is important that the ATE can test relatively small parts of complex circuits separately, to maximise test coverage, to enable effective fault diagnosis and to minimise test time.

These requirements can be achieved only if the circuit to be tested is designed to include them. If the circuit is designed only in relation to its operating functions test coverage will be low, with adverse effects on quality, reliability and cost. Hardware design for test can be assisted by using *test synthesis* design software. This generates the DFT structures in the circuit to enable tests to be performed as listed above, as well as scan tests and self-test, as described later.

Since ICT is applied only at the level of loaded circuit boards, the component layout on the board is important. The edge of the board must be left clear, to provide a seal for the vacuum chuck to operate to bring the board down on to the probe array. Registration holes must be provided, to ensure that all of the test pins make accurate connection to all of the test points. Test points must be provided at all of the important circuit nodes, to enable the ATE to test each component in turn. On double-sided boards, large components should be mounted on the top, and all components that are mounted on the bottom must be carefully positioned to avoid preventing contact by the test pins. Circuit design features must be included where appropriate to assist the ATE, for example the ability to open feedback loops or to isolate components that cannot be controlled by the ATE, such as clocks.

Design for FT must enable the tester to perform the initialisation, control, observation and partition functions. Since the test inputs and outputs are via the circuit functional connector (e.g. board edge connector or box connector), there are practical limits to the extent to which this can be attained. When appropriate, additional connector pins can be provided to increase test coverage. Guided probes, which enable the ATE operator to make temporary connections to selected test nodes on the circuit as instructed by the program, can also be used.

The need for partitioning is caused by the exponential growth in test complexity with the increases in the numbers of test conditions in modern digital circuits, as described earlier. Partitioning the circuit enables near-100 percent test coverage within practicable test times.

8.5.2 Test software

Test software must be developed for any circuit that is to be tested on ATE. *Automatic test program generation* (ATPG) software such as LabVIEW™ (National Instruments Corp.) is available to perform this task (reference 4). The ATPG will create the test software and indicate the fault coverage. For some systems, particularly aerospace electronics (avionics) and military systems, the ATLAS (Abbreviated Test Language for All Systems) test software is used.

8.5.3 Scan design

For very large digital circuits, internal circuits are often provided which assist the ATE to perform the tests. The best known is *Level Sensitive Scan Design* (LSSD), or *internal scan*. Internal scan (full, almost-full and partial) is used to separate combinational from sequential circuitry and to break internal feedback loops to facilitate ATPG. It also provides for excellent circuit partitioning, control and observability. At the printed circuit board assembly level, the best known approach is *boundary scan*. A boundary scan circuit receives test instructions from the ATE, then performs logical tests on the interconnects between the digital circuits that it controls. This capability is important for large IC packages containing many connections, particularly if they cannot otherwise be accessed for test by the ATE. Boundary scan can also be used to test some on-chip circuit functions. The international standard for the boundary scan approach is IEEE1149.1. Most modern digital ICs include boundary scan capabilities, and it is important to include them in ASIC designs to facilitate board or other assembly-level testing. The boundary scan method is described in Reference 5.

8.5.4 Built-in self-test (BIST)

Built-in self-test (BIST) is the capability of a circuit to run tests on itself whilst operating as part of a larger system. BIST might be continuous or on demand, for example on switch-on, and is usually driven by the system software, so that both hardware and software design might be involved. We observe BIST in action whenever we turn on our PCs.

Design for test techniques are described in Reference 6.

8.6 ELECTRONIC COMPONENT TEST

Since electronic components are always tested by their manufacturers, and since modern manufacturing quality is very high, it is very seldom appropriate or necessary for components to be tested by equipment or system manufacturers. This was not always the case: up to about the early 1980s many system manufacturers either tested complex components such as integrated circuits, or had them tested by independent test organisations, prior to assembly into circuits. The rapidly improving quality levels, together with the very high cost of test equipment for these components, has practically eliminated this practice.

We can consider electronic components to be in two categories from the test point of view: 'discrete' components and integrated circuits (ICs).

8.6.1 'Discretes'

Components such as resistors, capacitors, connectors, coils, transistors and other relatively simple types generally perform a discrete function and possess a relatively small number of performance parameters. Therefore testing them is relatively easy, quick and inexpensive. The manufacturing processes for these components are also relatively easy to control, resulting in high yields, or low proportions defective. Since the tests are quick and inexpensive, and the market expectations for quality are very high (typically for fewer than one per million defective), 100 percent functional testing by their manufacturers is justified and is normal. Therefore testing by users is rarely cost-effective.

8.6.2 Integrated circuits

ATE for IC testing must be capable of testing all or most of the functions of the very complex circuits, at the maximum operating speeds. ICs are usually functionally tested at two stages in the production cycle. Individual circuits are tested, using a wafer probe, before the wafer is scribed and broken into separate chips. The circuits that pass are then packaged and retested, via the package connections. Obviously both of these provide very limited access in terms of the very large numbers of circuit nodes.

The other main feature that the ATE must possess is the ability to generate very large numbers of test patterns, and to observe the responses, in very short times. Ideally the data rates should be in excess of the maximum capability of the IC, so very high speed circuits are necessary (over 1 GHz). In turn this forces the need for the 'front end' ATE circuits to be located as close as practicable to the IC under test, to minimise the lengths of conductors. The ATE circuits require active liquid cooling to prevent overheating.

Figure 8.8 shows a production IC ATE system.

The problem of testing very large and fast ICs is one of the major challenges of modern electronics technology. The circuit designs are created using specialised EDA software, and this includes capabilities for testing the logical and timing aspects of the circuit function, as well as for automatic test program generation (ATPG). However, 100 percent coverage is not attainable in practical timescales, and the results of this were dramatically demonstrated by both the Intel 486 and

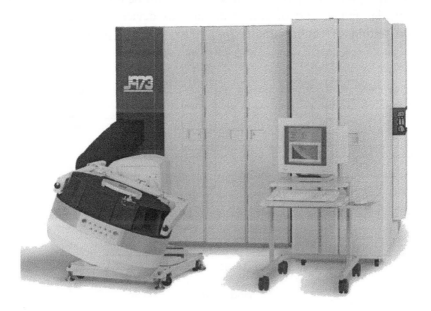

Figure 8.8 Integrated circuit ATE system (courtesy Teradyne)

Pentium microprocessors, which were put into production and initially marketed with logical design faults which had escaped all EDA analysis and hardware testing. Future generations of VLSI/ULSI, containing tens of millions of circuit elements and operating at GHz frequencies, will make the test problem even more challenging. Already the cost of testing such components is approaching the cost of manufacture.

New methods for testing very large and fast ICs are being developed, in response to the challenges of complexity and cost, principally I_{DDQ} testing and *built-in self-test* (BIST).

I_{DDQ} *testing*

I_{DDQ} is the current drawn by the device when it is powered but stable, i.e. not switching, or *quiescent*. The I_{DDQ} value at the nominal supply voltage level is typically very low, of the order of microamperes or nanoamperes. In the I_{DDQ} test, the supply voltage is set initially at the nominal V_{DD} level, and is then changed in steps to about six different levels. I_{DDQ} is measured at a number of logic (node) states at each V_{DD} level. Characteristic responses for good devices at a particular V_{DD} level might be as in Figure 8.9. If, however, a device is faulty in some way (a capacitor shorted, a transistor having a low gain, a track open circuit, etc.) the I_{DDQ} plot may be anomalous, for example as in the dotted lines on Figure 8.9. The plot does not indicate the actual fault, nor how it might affect function, but it does show that something is wrong. With experience, different anomalous I_{DDQ} plots for particular device types can be interpreted to correlate with particular faults, and of course anomalous devices can be subjected to functional testing to confirm these. I_{DDQ} tests are also performed using the

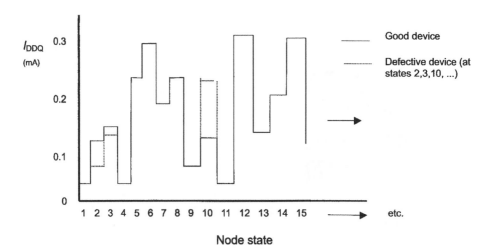

Figure 8.9 I_{DDQ} plot

normal supply voltage level, while applying a small number of test vectors and measuring the current at each. I_{DDQ} capabilities are built in to some ATE systems.

I_{DDQ} testing has some important advantages over functional testing:

- It can be performed with simple test circuits, software and interfaces. Only supply voltage, a number of test vectors (typically about 100 for sufficiently high fault coverage) and measurement of I_{DDQ} are required. However, very low current values must be accurately measured, and this requires special instrumentation and time. In particular, the I_{DDQ} values for very complex modern ICs are very low (nanoamps) and therefore difficult to measure accurately and quickly.

- Some faults can be found only, or more readily, with I_{DDQ} tests, particularly high-resistance shorts.

- Because of the simplicity, I_{DDQ} testing is replacing burn-in in some IC manufacturing situations.

However, I_{DDQ} testing cannot fully replace functional testing, since a 'good' result does not necessarily imply that all functions will be correct. Also, the time required to make accurate measurements of very low current values is a problem with high rates of production.

Built-in self-test (BIST)

BIST circuitry is integrated into the chip design, and is then initiated and controlled by the ATE. BIST performs tests and reports the results, rather than merely facilitating the ATE to do this. BIST is becoming more common on large IC designs, as test access becomes more difficult. A major advantage of BIST is that the IC retains its test capability, even when embedded into a system. It also makes possible tests that might otherwise not be feasible using external ATE, such as at-speed tests on internal functions.

8.6.3   IC design for test

IC designers must use the ATPG facilities of the EDA to provide the test vectors that will be used. They must also ensure that test coverage is maximised, and that critical functions are identified and given special attention. The methods described above for circuit design for test are equally applicable to IC design, particularly boundary scan and BIST. Reference 7 covers these topics.

The economics of component testing are covered in Chapter 10.

8.7   EMI/EMC TESTING

As described in Chapter 3, electromagnetic interference and electromagnetic compatibility (EMI/EMC) are very important factors in all modern electronic system design. It is nearly always necessary to test new designs for these aspects. However, once the design has been proven it is not usually necessary to perform EMI/EMC tests on manufactured units.

EMI testing measures the electrical or electromagnetic 'noise' outputs of equipment. Electrical noise outputs can contaminate the power supply which the equipment might share with other systems, or outputs that connect to other systems. Emitted electromagnetic noise might interfere with other systems, or with functions within the system.

Electromagnetic compatibility is the capability of the system to operate correctly in the presence of known levels of conducted or radiated noise.

Testing for EMI/EMC consists of operating the equipment and measuring the noise outputs, and submitting the equipment to the specified conditions. For radiated emissions and compatibility testing *anechoic chambers*, equipped with appropriate measuring and transmitting equipment, are used.

The requirements for EMI/EMC testing for approval of new designs are described in Chapter 13. References 8 and 9 describe EMI/EMC requirements and test methods.

8.8   INFORMATION

Information on suppliers of the equipment and services described in this chapter is provided in the book homepage (Preface page xvii)

REFERENCES

1. Bennetts, R. G., 1989, Design of Testable Logic Circuits, Addison-Wesley.
2. Brindley, K., 1991, *Automatic Test Equipment*, Newnes.
3. GenRad, Inc., *Managing Your Manufacturing Process: Focus on In-Circuit Testing* (see Appendix 3).
4. Johnson, G. W., 1997, *LabVIEW Graphical Programming* (2nd edn), McGraw-Hill.
5. Parker K. P., 1998, *The Boundary-Scan Handbook* (2nd edn), Kluwer Academic.
6. Turino, J., 1990, *Design to Test* (2nd edn), Van Nostrand Reinhold.
7. F. P. M. Beenker, R. G. Bennetts, A. P. Thijssen, 1997, *Testability Concepts for Digital ICs*, Kluwer Academic Press.
8. Ott, H. W., 1988, *Noise Reduction Techniques in Electronic Systems* (2nd edn), Wiley-Interscience.
9. Williams, T., 1992, *EMC for Product Designers: Meeting the EMC Directive*, Newnes.

9

Software[1]

9.1 INTRODUCTION

Software is now part of the operating system of a very wide range of products and systems, and this trend continues to accelerate with the opportunities presented by low-cost microcontroller devices. Software is relatively inexpensive to develop, costs very little to copy, weighs nothing, and does not fail in the ways that hardware does. Software also enables greater functionality to be provided than would otherwise be feasible or economic. Performing functions with software leads to less complex, cheaper, lighter and more robust systems. Therefore software is used increasingly to perform functions that otherwise would be performed by hardware, and even by humans. Recent examples are aircraft flight control systems, robotic welders, engine control systems, domestic bread-making machines, etc.

The software 'technology' used today is the same basic sequential digital logic first applied in the earliest computers. The only significant changes have been in the speed and word length capability of processors and the amount of memory available, which in turn have enabled the development of high-level computer languages and modern operating systems. Some attempts have been made to develop radically different approaches such as parallel processing and fuzzy logic, but these remain fringe applications. Therefore the basic principles of software development, to ensure that programs are correct, safe and reliable, remain largely unchanged since they were first described in the 1970s (References 2 and 3).

Every copy of a computer program is identical to the original, so failures due to variability cannot occur. Also, software does not degrade, except in a few special senses,[2] and when it does it is easy to restore it to its original standard.

---

[1] This chapter is based in part on Chapter 10 of *Practical Reliability Engineering* (3rd edition), John Wiley & Sons (1995) (Reference 1).

[2] Data or programs stored in some media can degrade. Magnetic media such as discs and tape are susceptible to electromagnetic fields, or even to being closely packed for long periods. VLSI semiconductor devices can suffer changes in the voltage state of individual memory cells due to naturally occurring alpha-particle bombardment. In each case a refresh cycle will restore the program.

Therefore a correct program will run indefinitely without failure, and so will all copies of it. However, software can fail to perform the function intended, due to undetected errors. When a software error does exist, it exists in all copies of the program, and if it is such as to cause failure in certain circumstances, the program will always fail when those circumstances occur.

Software failures can also occur as a function of the machine environment, e.g. machines can be restarted and the software 'fixed' by clearing queues, removing memory leaks, and refreshing the state of the machine. So identical copies can behave differently depending on their 'age' since rebooting.

Software errors can cause system failure effects that can range from trivial to catastrophic. Therefore software reliability and safety effort must be directed at the consequences of errors, not just at the prevention or removal of most of them.

Since most programs consist of very many individual statements and logical paths, all created by the efforts of humans, there is ample scope for errors. Therefore we must try to prevent the creation of errors, and maximise the likelihood of detecting and correcting those that are created, by imposing programming disciplines, by checking and by testing.

When software is an integral part of a hardware–software system, system failures might be caused by hardware failures or by software errors. When humans are also part of the system they can also cause failures (e.g. the Airbus crash during a low flying display, which some 'experts' immediately blamed on the new flight control software, but which the investigation concluded was caused by the pilot putting the aircraft into a situation from which the system could not prevent the crash). In some cases it might be difficult to distinguish between hardware, software and human causes.

There are several ways by which hardware and software reliability differ. Some have already been mentioned. Table 9.1 lists the differences.

9.2   SOFTWARE IN ENGINEERING SYSTEMS

The software that forms an integral part or sub-system of an engineering system is in some important ways different from software in other applications, such as banking, airline booking, logistics, CAE, PC operating systems and applications, etc. The differences are as follows.

- Engineering programs are 'real time': they must operate in the system time-scale, as determined by the system clock and as constrained by signal propagation and other delays (switches, actuators, etc.). A chess program or a circuit simulation program, for example, will run when executed, and it is not critical exactly how long it takes to complete the run. However, in an operational system such as a process controller or an autopilot, it is essential that the software is ready to accept inputs and completes tasks at the right times. The software must be designed so that functions are correctly timed in

Table 9.1 Comparison of Hardware and Software Reliability Characteristics

Hardware	Software
1 Failures can be caused by deficiencies in design, production, use and maintenance.	Failures are primarily due to design faults. Repairs are made by modifying the design to make it robust against the condition that triggered the failure.
2 Failures can be due to wear or other energy-related phenomena. Sometimes warning is available before failure occurs (e.g. system can become noisy, indicating degradation and impending failure).	There are no wearout phenomena. Software failures occur without warning, although very old code can exhibit an increasing failure rate as a function of errors introduced into the code while making functional code upgrades.
3 No two items are identical. Failures can be caused by variation.	There is no variation: all copies of a program are identical.
4 Repairs can be made to make equipment more reliable. This would be the case with preventive maintenance where a component is restored to an as-new condition.	There is no repair. The only solution is redesign (reprogramming), which, if it removes the error and introduces no others, will result in higher reliability.
5 Reliability can depend on burn-in or wear out phenomena; that is, failure rates can be decreasing, constant or increasing with respect to time.	Reliability is not so time-dependent. Reliability improvement over time may be affected, but this is not an operational time relationship. Rather, it is a function of reliability growth of the code through detecting and correcting errors.
6 Reliability may be time-related, with failures occurring as a function of operating (or storage) time, cycles, etc.	Reliability is not time related. Failures occur when a specific program step or path is executed or a specific input condition is encountered, which triggers a failure.
7 Reliability may be related to environmental factors (temperature, vibration, humidity, etc.).	The external environment does not affect reliability except insofar as it might affect program inputs. However, the program reliability is a function of the internal machine environment (queues, memory leakage, etc.).
8 Reliability can be predicted, in principle but mostly with large uncertainty, from knowledge of design, parts, usage, and environmental stress factors.	Reliability cannot be predicted from any physical bases, since it entirely depends on human factors in design. Some approaches exist based on the development process used and the extent of the code, but these are controversial.

(*continued*)

Table 9.1 (*continued*)

Hardware	Software
9 Reliability can be improved by redundancy. The successful use of redundancy presumes ready detection, isolation, and switching of assets.	Reliability cannot be improved by redundancy if the parallel paths are identical, since if one path fails, the other will have the error. It is possible to provide redundancy by having diverse parallel paths with different programs written by different teams.
10 Failures can occur in components of a system in a pattern that is, to some extent, predictable from the stresses on the components and other factors. Reliability critical lists are useful to identify high-risk items.	Failures are rarely predictable from analyses of separate statements. Errors are likely to exist randomly throughout the program, and any statement may be in error. Most errors lie on the boundary of the program or in its exception handling. Reliability critical lists are not appropriate.
11 Hardware interfaces are visual; one can see a 10-pin connector.	Software interfaces are conceptual rather than visual.
12 Computer-aided design systems exist that can be used to create and analyse designs.	There are no computerised methods for software design and analysis. Software design is more of an 'art form' lacking the provability of hardware, except to a limited extent through *formal methods* (see later).
13 Hardware products use standard components as basic building blocks.	There are no standard parts in software, although there are standardised logic structures. Software reuse is being deployed, but on a limited basis.

relation to the system clock pulses, task execution times, interrupts, etc. Timing errors are a common cause of failure in real-time systems, particularly during development. They are often difficult to detect, particularly by inspection of code. Timing errors can be caused by hardware faults or by interface problems. However, logic test instruments (logic analysers) can be used to show exactly when and under what conditions system timing errors occur, so that the causes can be pinpointed.

- Engineering programs share a wider range of interfaces with the system hardware. In addition to basic items such as processors, memory, displays and keyboards, other engineering interfaces can include measurement sensors, A/D and D/A converters, signal analysers, switches, connectors, etc.

- Engineering programs might be 'embedded' at different levels within a system: for example the main operating program might be loaded and run from disc, tape or accessible PROM devices, but other software might be embedded in components which are less accessible, such as ASICs, programmable gate arrays, signal processing ICs and flash memory devices. (The BIOS chip in a PC is also an example of software embedded in this way).

- There is often scope for alternative solutions to design problems, involving decisions on which tasks should be performed by hardware (or humans) and which by software.

- Engineering software must sometimes work in electrically 'noisy' environments, so that data might be corrupted.

- Engineering programs are generally, though not always, rather smaller and simpler than most other applications.

Therefore it is very important that engineering software is developed (specified, designed, programmed, tested, managed) in close integration with the hardware and overall system work. The not uncommon practice of writing a software specification, then 'outsourcing' the program development work, should not be an option for important engineering software.

9.3 SOFTWARE ERRORS

Software errors ('bugs') can arise from the specification, the software system design and the coding process.

9.3.1 Specification errors

Typically more than half the errors recorded during software development originate in the specification. Since software is not perceivable in a physical sense, there is little scope for commonsense interpretation of ambiguities, inconsistencies or incomplete statements. Therefore software specifications must be very carefully developed and reviewed. The software specification must describe fully and accurately the requirements of the program. The program must reflect the requirements exactly. There are no safety margins in software design as in hardware design. For example, if the requirement is to measure $9\,V \pm 0.5\,V$ and to indicate if the voltage is outside these tolerances, the program will do precisely that. If the specification was incorrectly formulated, e.g. if the tolerances were not stated, the out-of-tolerance voltage would be indicated at this point every time the measured voltage varied by a detectable amount from $9\,V$, whether or not the tolerances were exceeded. Depending upon the circumstances this might be an easily detectable error, or it might lead to unnecessary checks and adjustments because the out-of-tolerance indication is believed. This is a relatively simple example. Much more serious errors, such as a misunderstanding or omission of the logical requirement of the program, can be written into the specification. This type of error can be much harder to correct, involving considerable reprogramming, and can be much more serious in effect.

The Eurospace Ariane 5 spacecraft launcher failure was caused by such an error: the guidance computer and the inertial measurement unit used different bit formats for numerical data, but, even though this fact was known, no compensation was made because it had not resulted in failures on previous Ariane launchers. The new launcher's greater rocket thrust led to an overflow when the inertial unit measured velocities higher than experienced before. The NASA Mars Polar Orbiter spacecraft collided with the planet because part of the system was designed using measurements in miles while an interfacing subsystem used kilometres.

9.3.2   Software system design

The software system design follows from the specification. The system design may be a flowchart and would define the program structure, test points, limits, etc. Errors can occur as a result of incorrect interpretation of the specification, or incomplete or incorrect logic. Errors can also occur if the software cannot handle data inputs that are incorrect but possible, such as missing or incorrect bits.

An important reliability feature of software system design is *robustness*, the term used to describe the capability of a program to withstand error conditions without serious effect, such as becoming locked in a loop or 'crashing'. The robustness of the program will depend upon the design, since it is at this stage that the paths to be taken by the program under error conditions are determined.

9.3.3   Software code generation

Code generation is a prime source of errors, since a typical program involves a large number of code statements. Typical errors can be:

- Typographical errrors (did you spot it?)

- Incorrect numerical values, e.g. 0.1 for 0.01

- Omission of symbols, e.g. parentheses

- Inclusion of variables which are not declared, or not initialised at the start of program run

- Inclusion of expressions which can become indeterminate, such as division by a value which can become zero

- Accidental shared use of memory locations.

Changes to code can have dire consequences. The likelihood of injecting new faults can run as high as 50%, and tends to be highest for small code changes. The injected faults tend to be more obscure and harder to detect and remove. Changes can be in conflict with the original architecture and increase code complexity.

We will briefly describe the methods that can be used to minimise the creation of errors, and to detect errors that might have been created.

9.4 PREVENTING ERRORS

9.4.1 Specification

The overall system specification and the software specification must be prepared in harmony. Both should allow flexibility in relation to allocation of functions and should encourage integrated engineering.

Software specifications should be more than just descriptions of requirements. They must describe the functions to be performed, in full and unambiguous detail, and the operating environment (hardware, memory allocation, timing, etc.). They should also describe explicitly all of the conditions that must *not* be allowed to occur. They should describe the program structure to be used, the program test requirements and documentation needed during development, as well as basic requirements such as the programming language, memory alloca-tions, inputs and outputs. By adequately specifying these aspects, a framework for program generation will be created which minimises the possibilities for creating errors, and which ensures that errors will be found and corrected.

The specifications must be carefully reviewed, to ensure that they meet all of the requirements described above, and contain no ambiguities. Specification review must be performed by the project team, including the programmers and engineers whose work will be driven by the specifications.

9.4.2 Structure and modularity

Structured programming is an approach that constrains the programmers to using certain clear, well-defined approaches to program design, rather than allowing total freedom to design 'clever' programs which might be complex, difficult to understand or inspect, and prone to error. Structured programming leads to fewer errors, and to clearer software that is easier to check and maintain. On the other hand, structured programs might be less efficient in terms of speed or memory requirements.

Modular programming breaks the program requirement down into separate, smaller program requirements, or modules, each of which can be separately specified, written and tested. Each module specification must state how the module is to interface with other parts of the program. Thus, all the inputs and outputs must be specified. The overall problem is thus made easier to understand and this is a very important factor in reducing the scope for error and for easing the task of checking. The separate modules can be written and tested in a shorter time, thus reducing the chances of changes of programmer in mid-stream.

Structured programming might involve more preparatory work in determining the program structure, and in writing module specifications and test requirements. However, like good groundwork in any development programme, this effort is likely to be more than repaid later by the reduced overall time spent on program writing and debugging, and it will also result in a program which is easier to understand and to change. The capability of a program to be modified fairly easily can be compared to the maintainability of hardware, and it is often a very important feature. Program changes are necessary when logical corrections have to be made, or when the requirements change, and there are not many software development projects in which these conditions do not arise.

The optimum size of a module depends upon the function of the module and is not solely determined by the number of program elements. The size will usually be determined to some extent by where convenient interfaces can be introduced. As a rule of thumb, modules should not normally exceed 100 separate statements or lines of code in a high-level language, and less in Assembler code.

9.4.3   Software reuse

Sometimes existing software, for example from a different or previous application, can be used, rather than having to write a new program or module. This approach can lead to savings in development costs and time, as well as reducing the possibility of creating new errors. However, be careful! Remember Ariane 5 and Polar Orbiter!

9.4.4   Programming style

Programming style is an expression used to cover the general approach to program design and coding. Structured and modular programming are aspects of style. Other aspects are, for example, the use of 'remark' statements in the listing to explain the program, 'defensive' programming in which routines are included to check for errors, and the use of simple and consistently applied constructs whenever practicable. A well-disciplined programming style can have a great influence on software reliability and maintainability, and it is therefore important that style is covered in software design guides and design reviews, and in programmer training. References 2 and 3 cover this aspect of software development.

9.4.5   Software checking

To confirm that the specification is satisfied, the program must be checked against each item of the specification. For example, if a specification calls for a voltage measurement of $15\,V \pm 1\,V$, only a line-by-line check of the program listing is likely to discover an error that calls for a measurement tolerance of $+1\,V$, $-0\,V$.

Program checking can be a tedious process, but it is made much easier if the program is structured into well-specified and understandable modules, so that an independent check can be performed quickly and comprehensively. Like hardware design review procedures, the cost of program checking is usually amply repaid by savings in development time at later stages.

Formal program checking, involving the design team and independent people, is called a *structured walkthrough*, or a *code review*.

9.4.6 'Formal' design and analysis methods

Several so-called 'formal' software design methods have been developed, with the objective of setting up a disciplined framework for specification and programming that will reduce the chances of errors being created. Included among these are the *Vienna Development Method* (VDM), developed by IBM.

Methods have also been developed for automatically checking programs. These check the program for consistency, structure and logical errors, by assessing the source code against the specification. This is called *static analysis*. They are still being developed, and their use is somewhat controversial, since they set out to prove logically and mathematically that the specification meets the requirements and that the program is logically correct. Properly used, they can lead to better specifications and program designs, and can show up errors. However, they are often expensive to apply, and cannot provide total assurance against human fallibility.

9.4.7 Fault tolerance

Programs should be written so that errors do not cause serious problems or complete failure of the program or system. A program should be able to find its way gracefully out of an error condition and indicate the error source.

Software can also provide back-up or safety features to cater for hardware failures. This can be achieved by programming internal tests, with a reset and error indication if the set conditions are not met. Where safety is a factor, it is important that the program sets up safe conditions when an error or failure occurs. Examples of this approach are as follows.

- Checks of cycle time for a process (e.g. time to fill a tank), and automatic shutdown if the correct time is exceeded by a set amount. This might be caused by failure of a sensor or a pump, or by a leak.

- Failure of a thermostat to switch off a heating supply can be precluded by ensuring that the supply will not remain on for more than a set period, regardless of the thermostat output.

- Checks for rates of change of input values. If a value changes by more than a predetermined amount, take corrective action as above. For example, a pressure measurement might abruptly change to zero because of a transducer or connector failure, but such an actual pressure change might be impossible. The system should not be capable of inappropriate response to a spurious input.

- Allowance of two or more program cycles for receipt of input data, to allow for possible data loss, interruption or corruption.

Features such as these can be provided much more easily with software than with hardware, at no extra material cost or weight, and therefore the possibility of increasing the reliability and safety of software-controlled systems should always be analysed in the specification and design stages. Their provision and optimisation is much more likely when the software development is managed as part of an integrated, system approach.

9.4.8 Redundancy/diversity

Fault tolerance can also be provided by program redundancy. For high-integrity systems separately coded programs can be arranged to run simultaneously on separate but connected controllers, or in a time-sharing mode on one controller. A voting or selection routine can be used to select the output to be used. This approach is also called *program diversity*.

The effectiveness of this approach is based on the premise that two separately coded programs are very unlikely to contain the same coding errors, but of course this would not provide protection against a specification error. Redundancy can also be provided within a program by arranging that critical outputs are checked by one routine, and if the correct conditions are not present then they are checked by a different routine.

9.5 LANGUAGES

The selection of the computer language to be used can affect the reliability of software. There are three main approaches that can be used:

1. Machine code programming

2. Assembly level programming

3. High level (or high order) language (HLL or HOL) programming.

Machine code programming is the creation of the microcode that the processor runs. However, programming at this level should not be used, since it confers no advantages in speed or memory, is very prone to creation of errors, is extremely difficult to check, and has no error trap capabilities.

Assembly level programs are faster to run and require less memory than HLLs. Therefore they can be attractive for real-time systems. However, assembly level programming is much more difficult and is much harder to check and to modify than HLLs. Several types of error which can be made in assembly level programming cannot be made, or are much less likely to be made, in a HLL. Therefore assembly level programming is not favoured for relatively large programs, though it might be used for modules in order to increase speed and to reduce memory requirements. *Symbolic assemblers*, however, have some of the error-reduction features of HLLs.

Machine code and assembly programming are specific to a particular processor, since they are aimed directly at the architecture and operating system.

HLLs are processor-independent, working through a *compiler* which converts the HLL to that processor's operating system. Therefore HLLs require more memory (the compiler itself is a large program) and they run more slowly. However, it is much easier to program in HLLs, and the programs are much easier to inspect and correct. The older HLLs (FORTRAN, BASIC) do not encourage structured programming, but the more recently developed ones (Pascal, Ada, C, C++) do.

Since HLLs must work through a compiler, the reliability of the compiler can affect system reliability. Compilers for new HLLs and for new processors sometimes cause problems for the first few years until all errors are found and corrected. Generally speaking, though, compilers are reliable once fully developed, since they are so universally used. Modern compilers contain error detection, so that many logical, syntactical or other errors in the HLL program are displayed to the programmer, allowing them to be corrected before an attempt is made to load or run it. Automatic error correction is also possible in some cases, but this is limited to certain specific types of error.

Fuzzy logic is used to a limited extent in some modern systems. The ways in which fuzzy logic programs can fail are basically the same as for conventional logic.

Programmable logic controllers (PLCs) are often used in place of processors, for systems such as machine tools, factory automation, train door controls, etc. Programming of PLCs is much easier than for microprocessors, since only basic logic commands need to be created. PLC-based systems also avoid the need for the other requirements of processor-based systems, such as operating system software, memory, etc., so they can be simpler and more robust, and easier to test.

9.6   ANALYSIS OF SOFTWARE DESIGN

It is not practicable to perform a failure modes and effects analysis (FMEA) (see Chapter 5) on software, since software 'components' do not fail. The nearest equivalent to a FMEA is a code review, but whenever an error is detected it is corrected so the error source is eliminated. With hardware, however, we cannot eliminate the possibility of, say, a transistor failure.

In performing a FMEA of an engineering system that combines hardware and software it is necessary to consider the hardware failure effects in the context of the operating software, since system behaviour in the event of a hardware failure might be affected by the software, as described above. This is particularly the case in systems utilising built-in-test software, or when the software is involved in functions such as switching redundancy, displays, warnings and shut-down.

The sneak analysis (SA) method described in Chapter 5 for evaluating circuit conditions that can lead to system failure is also applicable to software. Since a section of code does not fail but performs the programmed functions whether or not they are the intended ones, there is an analogy with an erroneous circuit design.

Software sneak conditions are:

1. Sneak output. The wrong output is generated.

2. Sneak inhibit. Undesired inhibit of an input or output.

3. Sneak timing. The wrong output is generated because of its timing or incorrect input timing or sequence.

4. Sneak message. A program message incorrectly reports the state of the system.

Reference 1 provides more details of the application of SA to software.

9.7   DATA RELIABILITY

Data reliability (or information integrity) is an important aspect of the reliability of software-based systems. When digitally coded data is transmitted, there are two sources of degradation:

1. The data might not be processed in time, so that processing errors are generated. This can arise, for example, if data arrives at a processing point (a 'server', e.g. a microprocessor or a memory address decoder) at a higher rate than the server can process.

2. The data might be corrupted in transmission or in memory by digital bits being lost or inverted, or by spurious bits being added. This can happen if there is noise in the transmission system, e.g. from electromagnetic interference or defects in memory.

System design to eliminate or reduce the incidence of failures due to processing time errors involves the use of queuing theory, applied to the expected rate and pattern of information input, the number and speed of the 'servers', and the queuing disciplines (e.g. first-in first-out (FIFO), last-in first-out (LIFO), etc.). Also, a form of redundancy is used, in which processed data is accepted as being valid only if it is repeated identically at least twice, say, in three cycles. This might involve some reduction in system processing or operating speed.

Data corruption due to transmission or memory defects is checked for and corrected using error detection and correction codes. The simplest and probably best known is the parity bit. An extra bit is added to each data word, so that there will always be an even (or odd) number of ones (even (or odd) parity). If an odd number of ones occurs in a word, the word will be rejected or ignored. More complex error detection codes, which provide coverage over a larger proportion of possible errors and which also correct errors, are also used. Examples of these are Hamming codes and BCH codes.

Ensuring reliable data transmission involves trade-offs in memory allocation and operating speed.

9.8   SOFTWARE TESTING

The objectives of software testing are to ensure that the system complies with the requirements and to detect remaining errors. Testing that a program will operate correctly over the range of system conditions is an essential part of the software and system development process. Software testing must be planned and executed in a disciplined way since, even with the most careful design effort, errors are likely to remain in any reasonably large program, due to the impracticability of finding all errors by checking, as described above. Some features, such as timing, overflow conditions and module interfacing, are not easy to check.

Few programs run perfectly the first time they are tested. The scope for error is so large, due to the difficulty that the human mind has in setting up perfectly logical structures, that it is normal for some time to be spent debugging a new program until all of the basic errors are eliminated.

There are limitations to software testing. It is not practicable to test exhaustively a reasonably complex program. The total number of possible paths through a program with n branches and loops is 2^n, analogous to the digital circuit testing problem discussed in Chapter 8. However, there is no 'ATE' for software, so all tests must be set up, run and monitored manually. It is not normally practicable to plan a test strategy which will provide high theoretical error coverage, and the test time would be exorbitant. Therefore the tests to be performed must be selected carefully to verify correct operation under the likely range of operating and input conditions, whilst being economical.

The software test process should be iterative, whilst code is being produced. Code should be tested as soon as it is written, to ensure that errors can be corrected quickly by the programmer who wrote it, and it is easier to devise effective tests for smaller, well-specified sections of code than for large programs. The earliest testable code is usually at the module level. The detection and correction of errors is also much less expensive early in the development programme. As errors are corrected the software must be retested to confirm that the redesign has been effective and has not introduced any other errors. Later, when most or all modules have been written and tested, the complete program must be tested, errors corrected, and retested. Thus design and test proceed in steps, with test results being fed back to the programmers.

It is usual for programmers to test modules or small programs themselves. Given the specification and suitable instructions for conducting and reporting tests, they are usually in the best position to test their own work. Alternatively, or additionally, programmers might test one another's programs, so that an independent approach is taken. However, testing of larger sections or the whole program, involving the work of several programmers, should be managed by a person with system responsibility, though members of the programming team should be closely involved. This is called *integration testing*. Integration testing covers module interfaces, and should demonstrate compliance with the system specification.

The software tests must include:

- All requirements defined in the specification ('must do' and 'must not do' conditions)

- Operation at extreme conditions (timing, input parameter values and rates of change, memory utilisation)

- Ranges of possible input sequences

- Fault tolerance (error recovery).

Since it may not be practicable to test for the complete range of input conditions, it is important to test for the most critical ones and for combinations of these. Random input conditions, possibly developed from system simulation, should also be used when appropriate to provide further assurance that a wide range of inputs is covered.

Software can be tested at different levels:

- *White box* testing involves testing at the detailed structural level, for aspects such as data and control flow, memory allocation, look-ups, etc. It is performed in relation to modules or small system elements, to demonstrate correctness at these levels.

- *Verification* is the term sometimes used to cover all testing in a development or simulated environment, for example using a host or lab computer. Verification can include module and integration testing.

- *Validation* or *black box* testing covers testing in the real environment, including running on the target computer, using the operational input and output devices, other components and connections. Validation is applicable only to integration testing, and it covers the hardware/software interface aspects, as described earlier.

These terms are not defined absolutely, and other interpretations are also applied.

9.8.1 Managing software testing

Software testing must be managed as an integral part of the overall system test plan. It is essential to plan the software tests with full understanding of how software can fail and in relation to the interfaces with the system hardware. In the system test plan the software should be treated as a separate sub-system (verification) and as part of the overall system (validation).

The test specifications (for modules, integration/verification, validation) must state every test condition to be applied, and the test reports must indicate the result of each test.

Formal 100 percent error reporting should also be started at this stage, if it is not already in operation. Reporting of software errors is an important part of the overall program documentation. The person who discovers an error may not be the programmer or system designer, and therefore all errors, whether discovered during checking, testing or use, must be recorded with full details of operating conditions at the time. Obviously all errors must be corrected, and the action taken must also be reported (changes to specification, design, code, hardware, as appropriate). The relevant test must be repeated to ensure that the correction works, and that no other errors have been created.

Formal configuration control should be started when integration testing commences, to ensure that all changes are documented and that all program copies at the current version number are identical.

For validation and other system tests, failures caused by software should be reported and actioned as part of an integrated engineering approach, since the most appropriate solutions to problems could involve hardware or software changes.

Each software test needs to be performed only once during development, unless there have been changes to the program or to related hardware. Of course there is no need to test software as part of the system production test process, since the software cannot vary from copy to copy or change over time.

There are some who argue that software testing (and checking) should be performed by people who are entirely independent of the programming team. This is called the '*cleanroom*' approach to software development. The approach is controversial, and is not consistent with the philosophy on which this book is based.

9.9 CONCLUSIONS

The versatility and economy offered by software control can lead to an under-estimation of the difficulty and cost of program generation. It is relatively easy to write a program to perform a simple defined function. To ensure that the program will operate satisfactorily under all conditions that might occur, and will be capable of being changed or corrected easily when necessary, requires an effort greater than that required for the basic design and first-program preparation. Careful groundwork of preparing and checking the specification, ensuring that hardware–software interfaces are thoroughly taken into account, planning the program structure and assessing the design and code against the specification are essential, or the resulting program could contain errors and will be difficult to correct. The cost and effort of debugging a large, unstructured program containing many errors can be so high that it is cheaper to scrap the whole program and start again.

Software that is reliable from the beginning will be cheaper and quicker to develop, so the emphasis must always be to minimise the possibilities of early errors and to eliminate errors before proceeding to the next phase. The essential elements of a software development project to ensure a safe and reliable product are:

1. Specify the requirements completely and in detail (system, software).

2. Make sure that all project staff understand the requirements.

3. Check the specifications thoroughly. Keep asking 'what if?'.

4. Design a structured program and specify each module fully.

5. Check the design and the module specification thoroughly against the system specification.

6. Check written programs for errors, line by line.

7. Plan module and system tests to cover important input combinations, particularly at extreme values.

8. Ensure full recording of all development notes, checks, tests, errors and program changes.

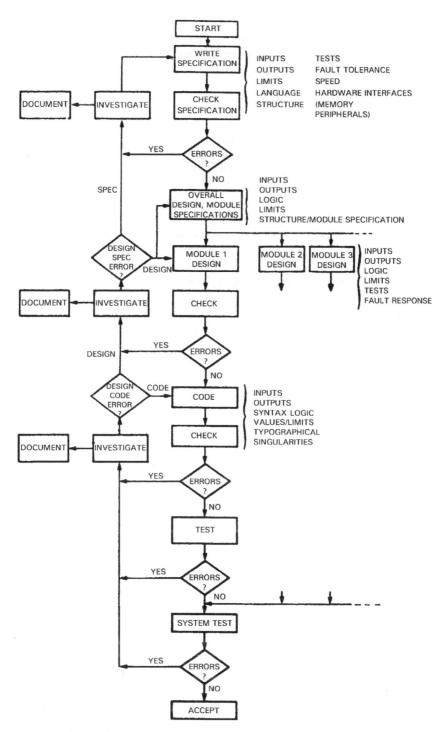

Figure 9.1 Software development for reliability and safety (reproduced from Reference 1 by permission of John Wiley & Sons Ltd)

Figure 9.1 shows the sequence of development activities for a software project, with the emphasis on reliability and safety. References 4–7 provide more information on software testing. References 8 and 9 provide excellent overviews of software reliability and safety engineering.

REFERENCES

1. O'Connor, P. D. T., 1995, *Practical Reliability Engineering* (3rd edn), John Wiley & Sons, Ltd.
2. Kernighan, B. W. and Plauger, P. J., 1974, *The Elements of Programming Style*, McGraw-Hill.
3. Yourdon, E., 1975, *Techniques of Program Structure and Design*, Prentice-Hall.
4. Myers, G. J., 1979, *The Art of Software Testing*, Wiley.
5. Ould, M. A. and Unwin, C., 1986, *Testing in Software Development*, Cambridge University Press.
6. Beizer, B., 1995, *Black-Box Testing*, John Wiley & Sons, Ltd.
7. Kaner, C., Quoc Nguyen, H. and Falk, J., 1999, *Testing Computer Software* (2nd edn), John Wiley & Sons, Ltd.
8. Musa, J. D., 1999, *Software Reliability Engineering*, McGraw- Hill.
9. Leveson, N. G., 1995, *Safeware—System Safety and Computers*, Addison Wesley.

10

Manufacturing Test

10.1 INTRODUCTION

All manufactured items should be correctly made, with all processes and parts being within specified tolerances. Manufacturing quality control and improvement techniques, as described in books such as References 1–3, are used to try to ensure this. However, such a controlled state can seldom be guaranteed, and manufacturing errors can occur. There are two kinds of defective item that can be produced as a result:

1. Those that do not work initially, or that can otherwise be identified as defective by inspection, measurement, etc.

2. Those that pass initial inspections and tests but contain defects that could cause failures later. Examples are:

   - Voids or cracks in castings, forgings or welds
   - Fasteners with incorrect torque loading
   - Inadequate protective coatings (plating, paint, etc.)
   - Inadequate solder joints
   - Weak capacitor dielectrics in integrated circuits
   - Electronic circuits that contain components with out-of-tolerance parameters in relation to stresses such as high or low temperatures.

We must therefore determine the inspection and test methods that will most effectively and economically identify these. Generally, 'inspection' is the term applied to checks, measurements, etc. made while the item is not operating, and 'test' refers to the item being in an operating state. However, these are not exclusive meanings.

In this chapter we will describe the main principles and methods for testing mechanical and electronic products in manufacture and the economics of manufacturing test. Technology aspects of mechanical and electronics test were covered in earlier chapters.

10.2   MANUFACTURING TEST PRINCIPLES

10.2.1   Value added testing

The first principle of testing manufactured products is that all testing adds to manufacturing costs, and therefore should be reduced to the minimum that is necessary to minimise the downstream costs of defects and failures that would otherwise arise. The ideal minimum is no testing, and this is sometimes realistic and practical. If all of the production processes are in control then no testing should be necessary, because we would know that only good items would be produced. However, for most engineering products and systems, testing is an essential element of manufacturing, because:

1. Manufacturing processes generate variation and defects.

2. The costs of defects later in the product life cycle (later manufacturing stages, in service) might exceed the costs of the tests and rectification during manufacture that could prevent them.

Therefore the correct way to think about manufacturing test economics is to consider the total life cycle. This leads to the concept of *value-added testing.* Testing should be considered as a value-adding activity, like any other manufacturing process, since the objective is to create a better and more economic product. The tests that are performed must be justified economically. However, in contrast to determining the value added by other processes, this can be difficult, because:

1. Test costs arise in manufacture, whereas defect and failure costs mostly arise later.

2. The occurrence and costs of defects and failures cannot be predicted with certainty.

The large amount of uncertainty that is often associated with defects and failures in service and the costs they might generate mean that manufacturing test is also *insurance testing*, and the value of this protection must be taken into account. Process or supplier problems can very quickly generate high costs (warranty, recall, liability, reputation, fixing), so these possibilities should be considered when planning testing.

Testing is sometimes necessary to ensure accurate or correct performance, for example, a low-noise, wideband amplifier whose performance depends critically on small variations in stage gain and noise of individual transistors.

For some products tests as part of manufacturing are mandatory. These include items like transducers, flowmeters, instruments, etc., which must be calibrated to provide evidence of accuracy, and large systems such as trains, aircraft, power stations, etc., for which final tests must be performed to demonstrate satisfactory operation before acceptance into service. (Calibration is discussed in the next chapter.) Obviously value is added by such tests, since the product cannot be sold without them.

10.2.2 Test capability

The second principle is that *the tests must be designed to detect the defects that might be generated by the manufacturing processes*. These include the processes of suppliers, handling, transport, storage and any other inputs to the operation. This is *test capability*. In order to maximise test capability it is necessary to know the capabilities of the manufacturing processes and the kinds and frequencies of the defects they might generate. As stated in point 2 above, this is not easy or certain, and defect generation can also fluctuate widely. Defects and later failures can be caused by process variations, as discussed in Chapter 4, and by many other causes. These will be described later in this chapter in relation to the technology aspects.

Test capability is also influenced by the defect detection methods that can be applied. Defects can be detected visually using manual or automatic methods, or by their effects on product behaviour or on measured parameters.

Test capability also has a statistical aspect in relation to the measurement of variables. The measurement inaccuracy or tolerance must be less than that of the measured value. A 10 to 1 ratio between the maximum parameter variation and that of the measurement is typically required.

Other important aspects of test capability are the ability to correctly indicate, when appropriate, the location or cause of the defect or failure so that it can be rectified, and the ability not to falsely indicate defects or failures of good items.

10.2.3 Test criteria

The third principle must be that manufacturing tests *should not include tests of the design of the product or processes*. The adequacy of the design should have been confirmed by the development tests, as discussed in Chapter 6. This means that manufacturing tests should not be copies of the tests applied in development, but should be optimised in relation to the engineering and economics of manufacture and service. In many cases the methods and equipment are very different from those used in development. It is important that the information gained in development testing is applied to the planning and optimisation of manufacturing test, and that development testing includes the evaluation and optimisation of the manufacturing test methods, but the tests should be repeated or copied only to the extent appropriate to detecting manufacturing problems.

10.2.4 Test stresses

The fourth principle is that any tests performed must detect defects or failures generated by the manufacturing processes, *without damaging or reducing the useful life of good items. This is a fundamentally different requirement compared with that for development testing.* Development tests should show weaknesses in product and process designs by generating failures, so the stresses applied should be well in excess of the expected worst cases. Manufacturing tests must avoid damaging good items, so the stresses applied must be suitably modified. We will discuss stress testing for manufacturing later.

10.3 MANUFACTURING TEST ECONOMICS

The economic justification for decisions on testing must be based upon the principles described above, in particular that of value-added testing. Each case has to be considered on its own merits. Aspects that should be included are:

- Cost of the test (setup, operating, repairs to items that fail, etc.)
- Defects/failures that could be introduced by upstream processes and their likely frequencies of occurrence
- Likelihood of the test detecting defects and failures (test capability)
- Alternative methods of detection, and their costs and effectiveness (inspection, etc.)
- Methods to reduce or prevent the occurrence of defects and failures, and their costs and effectiveness
- Downstream costs and other consequences of undetected defects and failures (later manufacturing stages, in service)
- Whether 100 percent or sample testing is appropriate (sampling is discussed in Chapter 12)

For many types of product and manufacturing situation it is relatively easy to determine the best approach. Thus, for example, an item such as a mass-produced automotive gearbox would not normally be subjected to operating tests, since the setup and operating costs would be high in relation to the likely costs of failures. However, a system such as an aircraft would be fully tested during and at the end of assembly, since the many manual assembly processes present relatively high probabilities of failure, and the costs and safety consequences of these can be very high. However, some situations are less clear-cut, particularly in electronics manufacture.

Many items and systems undergo a series of manufacturing processes, with associated inspections and tests. The principle of minimising the overall cost of manufacturing and service must be applied collectively to all of these. In such cases defects or failures can generate costs at downstream manufacturing stages or after shipment, and in most cases the later in the overall process that these are detected the higher are the costs they incur. For example, during the manufacture of a car some components and subsystems will be inspected or tested before installation (usually by their suppliers before delivery to the line), then the final product will be inspected and maybe tested. A final inspection is usually performed by the dealer before delivery to the customer.

If inspections or tests are performed, it is usually best if they are done as soon after and as close to the previous manufacturing stage as can be arranged. This helps to ensure that action can be taken quickly to correct the causes of problems before they can lead to repeated occurrences. In principle, each problem that is detected should lead to instant corrective action, and this is an important feature of the *kaizen* approach to manufacturing quality improvement (Reference 3).

10.4 INSPECTION AND MEASUREMENT

Inspection and measurement are production activities that are obviously closely related to testing. The term 'inspection' is generally used to cover any activity that detects condition or properties (missing components, cracks, discoloration, leaks, etc.). Inspection is usually, though not always, performed without the item being operated. Measurement is any quantified inspection of properties (dimensions, weight, electrical parameters, etc.). Some processes overlap these working definitions. For example, a hydraulic actuator might be inspected for leaks while it is functionally tested, and electronic in-circuit test (ICT) can be considered as an inspection of the assembly through tests of the components.

We can often determine the condition of products by these methods, either as a substitute for or to supplement testing. Examples are:

- Inspection of metal castings for surface condition, cracks, etc.

- Measurement of the machined dimensions on components such as engine cylinder blocks

- Visual or automatic optical inspection of an electronic assembly before it is submitted to test. This can minimise the danger of the test causing damage, and can reduce the total test cost.

Other items cannot be tested on their own, but only as part of a system. Machined engine components such as cylinder blocks, cylinder heads and other machined components can only be tested as part of the built-up engine.

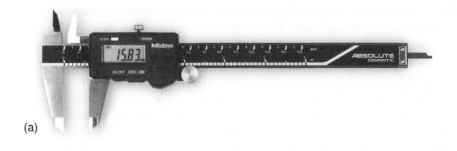

(a)

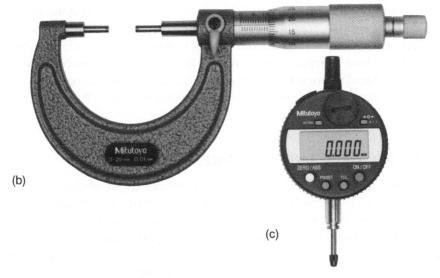

(b)

(c)

Figure 10.1 Manual dimensional measuring instruments: (a), caliper; (b) dial indicator; (c) micrometer (courtesy Mitutoyo)

However, it is important to know that they are dimensionally correct and within all tolerances before assembly and test.

The technology of dimensional measurement is called *metrology*. Dimensional measurements can be made using a variety of methods, including manual instruments such as calipers, micrometers and dial indicators (Figure 10.1), optical projection methods for measuring shapes such as gear teeth (Figure 10.2), and co-ordinate measuring machines (CMMs) which automatically measure and record a number of separate dimensions (Figure 10.3).

References 4–6 describe measurement methods in detail.

10.5   TEST METHODS

The methods to be used for manufacturing test (and inspection) should always be determined during design and as far as practicable optimised as part of the

Figure 10.2   Profile projector (courtesy Mitutoyo)

development test programme. If design analysis and development testing is planned and performed as described in Chapters 5 and 6, this will provide the knowledge of the parameters that should be monitored on test, the features that are most likely to cause problems or failures, the methods that are best suited, and the stresses that can safely be applied during test. Leaving the determination of test methods until later and failing to integrate the development and manufacturing test regimes is very likely to result in manufacturing tests which fail to provide the most cost-effective assurance that all defects and failures are detected, and therefore to higher in-service costs (warranty, support, reputation).

10.5.1   Mechanical and systems test

The kinds of defects that can occur in the manufacture of mechanical components and of systems are far too numerous to list. Some examples are:

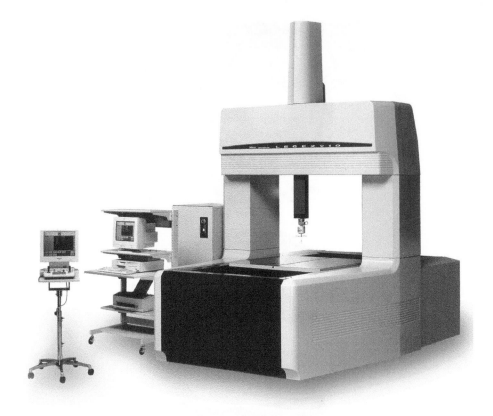

Figure 10.3 Co-ordinate measuring machine (CMM) (courtesy Mitutoyo)

- Any process variation that can exceed limits (machining, heat treatment, protective coating, etc.)

- Damage to parts during assembly, test, packing, transport, etc., particularly fragile or sensitive items like connectors, seals, etc.

- Incorrect fitting of parts (wrong part, wrong orientation, inadequate tightening, incorrect routing of pipes, cables, etc.)

- Fitting of defective (inoperative, out of tolerance) components or sub-assemblies.

Some of these can be detected and corrected easily and early, resulting in relatively low costs. Others can cause failures that occur later in service, and are therefore more serious.

The test methods described in Chapter 7 can all be applied as appropriate to manufacturing test. However, the condition and function of most simple mechanical components (fasteners, machined parts, bearings, etc.) can be determined without having to test them, by inspection or by control of the manufacturing processes. At the system level also functionality can often be assumed for many kinds of non-electronic assemblies, such as car bodies, aircraft wing structures and electric motors. At higher levels of assembly and for more complex systems it becomes progressively more appropriate to perform tests, since the likelihood of defects increases. For example, we would not normally test every component (pumps, valves, actuators, pipes, connectors, etc.) of a hydraulic or pneumatic system, but functional and leak tests are usually performed on assembled systems.

Most manufacturing testing of non-electronic systems can be performed without the need for complex or expensive test equipment. Correct performance can usually be determined by observation or measurement of a few parameters. We can test a car, an air-conditioning unit or a food mixer with little or no special equipment.

10.5.2   Electronics

The most frequent kinds of faults that typically occur in the manufacture of modern electronic assemblies and systems are:

- Defective solder connections (permanent or intermittent open or short-circuit, weak connection, non-adhesion due to non-wetting of surfaces, etc., as described in Chapter 3)

- Components missing, wrongly placed or wrong value

- Component parameters or functions incorrect (not working, out of tolerance)

- Component or other damage caused by handling, electrostatic discharge (ESD), etc.

- System-level faults, including cable and connector damage, tolerance effects, etc.

These are listed roughly in the order of their relative frequencies of occurrence. During the 1970s the list could have been reversed, with faulty components being the highest contributor and solder problems the lowest. The trend to the present situation reflects the combined effect of the great improvements in the quality of most components and the increase in number and reduction in size of solder connections and vias (solder-plated holes in circuit cards). It has also been influenced by the increased application of digital technology as opposed to analogue, so that the great majority of circuit functions are performed within ICs. The trend has in turn influenced the development and use of test equipment for manufacturing, as described in Chapter 8. Whereas early ATE systems were relatively simple functional testers which could detect operating and parametric faults, ATE for testing large, fast modern circuit boards and systems has become very complex and expensive. Therefore the emphasis today is on cheaper automatic inspection systems (AOI, X-ray (AXI), MDA, ICT, flying probe, as described in Chapter 8) applied to circuit boards, with functional tests being applied mainly to ICs (by their manufacturers) and at the complete assembly level. Inspection methods cannot ensure that the circuit or system will work, so it is nearly always necessary to perform some functional testing as well. The probability of a manufactured circuit or system not working correctly or failing in service due to process or component faults is usually sufficiently high that in most cases it is necessary to perform functional tests.

Electronic equipment can also be tested in production using manual testers, but this is economic only in low-volume operations. Manual testers are, however, often used for diagnosing the causes of faults indicated but not fully diagnosed by ATE.

Electronic circuit boards, assemblies and systems are, of course, generally repairable, so most faults that are detected by inspection and test are repaired, and the units retested.

A proportion of the faults mentioned above (solder problems, component parameters, damage) can escape detection by inspection and test, but cause failures later in service, as discussed in Chapters 2 and 3.

10.6 STRESS SCREENING

Stress screening, or *environmental stress screening* (ESS), is the application of stresses that will cause defective production items which pass other tests to fail on test, while not damaging or reducing the useful life of good ones. It is therefore a method for improving reliability and durability in service. Other terms are sometimes used for the process, the commonest being *burn-in*, particularly for electronic components and systems, for which the usually applied stresses are high temperature and electrical stress (current, voltage). The stress levels and durations to be applied must be determined in relation to the main failure-generating processes and the manufacturing processes that could cause weaknesses. Stress screening is a normally a 100 percent test, i.e. all manufactured items are tested. Stress screening is applied mainly to electronic components and assemblies, but it should be considered for non-electronic items also, for example precision mechanisms (temperature, vibration) and high-pressure tests for pneumatics and hydraulics to check for leaks or other weaknesses.

ESS guidelines have been developed for electronic components and systems. The US Navy has published guidelines (NAVMAT P-9492), and the US Department of Defense published MIL-STD-2164 (ESS Guidelines for Electronics Production), but these were inflexible and the stress levels specified were not severe (typically temperature cycling between 20°C and 60°C for 8 hours, with random or fixed frequency vibration in the range 20–2000 Hz, and the equipment not powered or monitored). The US Institute for Environmental Sciences and Technology (IEST) developed more detailed guidelines in 1990 (Environmental Stress Screening of Electronic Hardware (ESSEH)), to cover both development and manufacturing test. These recommended stress regimes similar to the military ESS guidelines. The details were based to a large extent on industry feedback of the perceived effectiveness of the methods that had been used up to the preparation of the guidelines, so they represented past experience, particularly of military equipment manufacture. They were not based upon the results of development testing on a product-by-product basis, or on any scientific study of technologies and failure mechanisms.

10.6.1   Highly accelerated stress screening

Highly accelerated stress screening (HASS) is an extension of the HALT principle, as described in Chapter 6, that makes use of very high combined stresses. No 'guidelines' are published to recommend particular stresses and durations. Instead, the stresses, cycles and durations are determined separately for each product (or group of similar products) by applying HALT during development. HALT shows up the product weak points, which are then strengthened as far as practicable so that failures will occur only well beyond the envelope of expected in-service combined stresses. The stresses and times that are then applied during HASS are higher than the operating limit, and extend into the lower tail of the distribution of the permanent failure limit.

It is essential that the equipment under test is operated and monitored throughout.

The stresses applied in HASS, like those applied in HALT, *are not designed to represent accelerated service conditions. They are designed to precipitate failures which would otherwise occur in service.* This is why they can be developed only by applying HALT in development, and why they must be specific to each product design. Because the stress levels are so high, *they cannot be applied safely to any design that has not been ruggedised through HALT.*

Obviously the determination of this stress–time combination cannot be exact, because of the uncertainty inherent in the distribution of strength. However, by exploring the product's behaviour on test, we can determine appropriate stress levels and durations. The durations will be short, since the stress application rates are very high and there is usually no benefit to be gained by, for example, operating at constant high or low temperatures for longer than it takes for them to stabilise. Also, only a few stress cycles will be necessary, typically one to four.

When stresses above the operating limit are applied, it will not be possible to perform functional tests. Therefore the stresses must then be reduced to levels below the operating limit. The functional test will then show which items have been caused to fail, and which have survived. The screening process is therefore

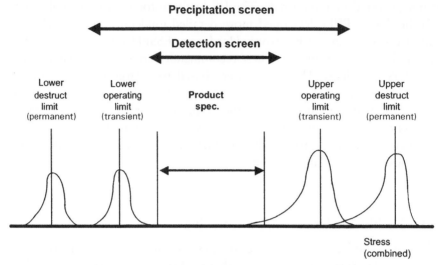

Figure 10.4   HASS philosophy (courtesy G. Hobbs)

in two stages: the *precipitation screen* followed by the *detection screen*, as shown in Figure 10.4.

Since the stresses applied in HASS can be very high, it is essential that they are determined with care, and validated before they are applied in production. The validation must ensure that the stresses will cause zero or negligible damage to good items, whilst showing up those that are weak. This stage of HASS development is called *proof* (or *safety*) *of screen*. A small sample of early production items must be subjected to the HASS profile determined as described above, then the screen is repeated a fairly large number of times (10 or more). This will show whether or not the HASS regime does reduce the useful life or otherwise damage good items. If it does, as evidenced by failure or observed damage, then the stresses must be re-evaluated. If no failures occur, consider running qualification tests to demonstrate that the items have not been damaged.

For items that are produced in large numbers, and when the manufacturing processes are considered to be under control, it is sometimes appropriate to screen only a sample, in order to detect whether batch-related problems (from suppliers or internally) or process trends are affecting quality. When applied to HASS this is referred to as *highly accelerated stress audit* (HASA). The sample size and frequency should be determined as described below for any sampling test or inspection.

The HASS development process is as follows:

1. Using the information gained during HALT on the operating (transient) and destruct (permanent) failure limits, select a HASS *precipitation screen* (stress(es) and durations) which is above the levels which would cause operating

failure (if possible), but below those which would damage or reduce the life of good items.

2. Select a HASS *detection screen*, which is below the operating limit but above the worst-case field stress levels.

3. Run *proof of screen*.

4. At the start of manufacturing, optimise the HASS using production hardware.

5. Depending on manufacturing quantities, experience and costs, consider moving to HASA.

The total test time for HASS is much less than for conventional ESS: *a few minutes compared with many hours*. HASS is also much more effective at detecting faults. *Therefore it is far more cost-effective, by orders of magnitude.*

HASS is applied using the same facilities, particularly environmental chambers, as are used for HALT. Since the test times are so short, it is often convenient to use the same facilities during development and in production, leading to further savings in relation to test and monitoring equipment, interfaces, etc.

The HASS concept can be applied to any type of product or technology. It is by no means limited to electronic assemblies. If the design can be improved by HALT, as described in Chapter 6, then in principle manufacturing quality can be improved by HASS. Since HASS is so much more effective than conventional ESS we will refer only to HASS in the discussions that follow.

HASS is described fully in Reference 7.

10.7 ELECTRONICS MANUFACTURING TEST OPTIONS AND ECONOMICS

10.7.1 Circuit board test

As described in Chapter 8, several options are available for inspection and testing of electronic circuit boards. The optimum policy for any product must be based upon the technologies being used, quantities to be made, and the other factors discussed earlier. Figure 10.5 shows a typical arrangement for manufacturing inspection and test flow of a circuit board. In this case AOI is the first inspection after assembly, and this is followed by MDA or ICT, then FT. All units that pass move on to the following stages, and are eventually shipped. Units that fail move to the diagnosis and repair station, after which they are re-submitted to MDA. The proportions that fail each stage are $d_{(x)}$. The costs of the assembly, inspection, test and diagnosis/repair operations (including amortised assembly and test equipment costs) are shown as $C_{(X)}$.

A simple model for the manufacturing and test cost per unit is:

$$C = C_A + C_I + C_M + C_F + (C_R + C_M + C_F)(d_i + d_m + d_f)$$

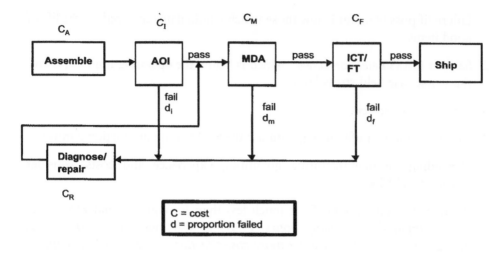

Figure 10.5   Test flow for an electronic assembly

If, for example,

$$C_A = \$200$$
$$C_I = \$10$$
$$C_M = \$10$$
$$C_F = \$20$$
$$C_R = \$50$$
$$d_i = d_m = d_f = 0.05$$

then the total cost per unit would be $252. The inspection, test and failure costs would add 26 percent to the basic manufacturing cost, and 23 percent of this would be due to failures. This is a fairly typical cost outcome, and it shows the high costs of test and of failures. The test costs reflect the high fixed cost of the test hardware, the costs of the test software and interfaces, and the variable costs of conducting the tests. Failure costs are high, and usually uncertain, because of the additional effort of diagnosis, repair, retest, documentation, corrective action, etc. Most diagnosis is performed by the ATE, but sometimes manual effort is also necessary.

The model described is simple and ignores some important practical aspects. Costs of faults usually increase if they are detected later in the process. The fault rates and test/inspection coverage values are very uncertain, depending on the components used, supply and process variations, testability of the circuits, test equipment and software, etc., and are also subject to unexpected changes. Also, every test has probabilities of correctly passing a good unit, incorrectly failing a good unit, and failing to detect a bad unit. This logic is shown in Figure 10.6, and

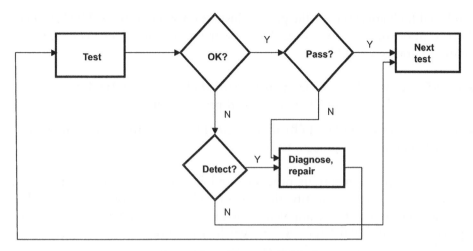

Figure 10.6 Test pass/fail logic

typical inspection and test coverage values against common faults are given in Table 10.1.

For the vast majority of circuits produced today, especially those produced by subcontract manufacturers, the largest cost is the functional testing of fault-free assemblies. With production yields of 85 percent to 95 percent, and with

Table 10.1 Electronic Manufacturing Fault Proportions and Inspection/Test Coverage

Fault	Faults	Coverage (%)				
		AOI	AXI	MDA/ICT	FT	HASS
Open circuit	25	40	95	85	95	*
Insufficient solder	18	40	80	0	0	20–80
Short-circuit	13	60	99	99	95	*
Component missing	12	90	99	85	85	*
Component misaligned	8	80	80	50	0	0
Component electrical parameter error	8	0	0	20/80	80	*
Wrong component	5	15	10	80	90	*
Other non-electrical	4	80	0	0	0	20–80
Excess solder	3	90	90	0	0	0
Component reversed	2	90	90	80	90	*

* Depends upon extent of FT performed concurrently.

combined fault coverage for the process inspection stages (vision, MDA/ICT) of 90 percent or so, very few process faults escape to functional test. For the above figures the range would be:

- for 85 percent yield (0.15 faults per board (FPB)) about 0.015 FPB escape to FT (1 in 63 boards fail FT with escaped process faults)

- for 95 percent yield (0.05 FPB) about 0.005 FPB escape to FT (1 in 200 boards fail FT with escaped process faults).

Therefore improvements in processes and in the effectiveness of upstream inspection methods can justify the use of reduced functional tests and less expensive functional test equipment. For example, full functional test can be replaced by a 'hot rig' (Chapter 8) into which production boards are inserted and tested by running some simple system tests.

The 'special to type' nature of many functional test systems, versus the universality of vision, MDA/ICT and flying probe testers means that capital costs of FT might be spread over fewer production units. (For MDA/ICT we must take account of fixture costs, as these are unique to each board.) Typical production lines may use several 'special to type' functional testers for each MDA/ICT at the first functional test stage. If there is a second functional test stage there will be more specialist testers. If the line produces several board types the number of functional testers will increase dramatically.

A further important issue can be the time required for the different tests, when high throughput rates are involved.

10.7.2   Assembly test

The model discussed above relates to a simple assembly consisting of only one board. For larger assemblies that typically might include several circuit boards, maybe involving different technologies (power, processing, microwave, etc.), cables and connectors, input devices such as keypads, display screens, enclosure, etc., there will be several test stages, using different types of test equipment. The test flow model would be accordingly more complex, and determining the optimum test approach presents further uncertainty. For example, should items such as power supplies be tested before being fitted, or as part of the final test of the assembly?

10.7.3   Integrating stress screening

If we are to make the most effective use of HASS it should be integrated into the test flow. Since units subjected to stress screens should always be operated and monitored, functional tests are being applied, so a further FT stage should not be necessary. Sometimes it might not be possible or practicable to perform the full range of functional tests during screening, and some further tests might be necessary.

Another aspect to be considered is the stages in the test flow at which stress screens should be applied. For example:

- Should stress screens be applied before or after AOI/ICT? Prior stressing can make some faults more likely to be detected by AOI/ICT. Alternatively, AOI/ICT before stress screening can pick up faults so that they can be repaired before units incur the cost of the screen, and the screen then looks for faults not detected by AOI/ICT.

- Should screening be applied at subassembly stages, or only after final assembly? Screens can be made more effective when designed, using HALT, at the level of subassemblies. On the other hand, screening at assembly level can detect faults generated during the final manufacturing stages. Each subassembly must be considered, in relation to aspects such as expected types and proportions of faults, size, test cost, etc.

- Should repaired units be screened again? The usual answer is 'yes'.

These are all difficult questions, and it is not always clear what the optimum approach should be. In particular, since HALT and HASS are relatively new concepts, the options must be considered carefully and the experience of others should be sought where practicable. See Reference 7.

10.7.4   Post-production savings

So far we have considered only the costs of inspection and test. However, we must relate these to the savings generated by preventing failures after shipping. If, in the example above, we performed no tests, and as a result 10 percent of shipped units were faulty, and if the cost of a failure in service was $1000, then the total cost of failures per unit would be $100, easily justifying the costs of test ($52). This assumes that all potential failures are detected by the tests (100 percent coverage). For complex, few-of-a-kind assemblies and systems it is nearly always easy to justify the cost of manufacturing tests in this way. However, for mass-produced items such as mobile phones, PC circuits, TV circuits, etc., it can be more difficult to justify tests that add 10–30 percent to manufacturing costs, and careful analysis and optimisation is important, to minimise both manufacturing and in-service costs.

The basic inspection and test methods discussed above apply only electrical stresses, and these are not accelerated. The tests are generally performed under benign conditions, and for very short durations. Therefore, they will show up only parametric and operating failures that are caused by some of the faults listed in Table 10.1. They will not show up faults that could cause failures under more severe operating conditions, or later in the life of the product. Also, they are unlikely to show up intermittent failures, such as shorts or opens that occur only under mechanical, vibration or thermal stress. HASS shows intermittents, stress-related faults, degradation/durability faults, etc. not covered by other tests. Therefore its economic justification should be made separately, based upon the expected occurrence and costs of these kinds of faults. For example, if the cost of failure in service is $1000, and 5 percent of units fail in service due to faults that are not detected by other tests, then the failure cost/unit will be $50. If HASS costs $40/unit, and has 80 percent coverage of faults that could cause failures, then the cost/unit (HASS + failures) would be $(40 + 50 × 0.2) = $50, or break-even.

HASS can seldom be justified for high-volume manufacturing, due to the time required to set up and run each test or batch of tests. HASA can be considered in these situations (see Chapter 6 and Reference 7).

10.7.5 Conclusions

Decisions on electronics test optimisation depend upon knowledge and assumptions about a range of factors. Analysis should take account of process yields, fault types, fault coverage, inspection and test costs, fault costs, production quantities, product mix, experience, improvement plans, etc., to identify aspects that are important for optimising the total process. All of these are subject to considerable uncertainty and change. Therefore it is not helpful to prescribe any generally 'best' approach. The optimum for any particular manufacturing and test situation can be determined only by careful consideration of all of the important factors, including the information gained during development testing.

Possible improvements should not be limited to the manufacturing and inspection/test processes, but should also include design and development aspects such as specifications, tolerances, design for test and protection.

Planning and optimising electronic equipment testing during manufacture is a complex and dynamic business, involving very high costs both to assure quality and of failure to do so. Because of this, and of the high capital costs of automatic assembly and test equipment, the manufacture and test of electronic assemblies is often out-sourced to specialist companies. It is essential that there is close co-operation during all product phases between the designers and the engineers involved in manufacture and test.

Reference 8 covers the economics of electronics test in detail.

10.8   TESTING ELECTRONIC COMPONENTS

A feature of manufactured electronic components (and systems) is that it is not possible to determine their functionality by any means other than testing.

Therefore, electronic components are tested as part of their manufacturing processes. Simple components like resistors and transistors, as well as complex ones like memories and microprocessors, are nearly always 100 percent tested to determine their accuracy and performance (i.e. every component produced is tested, rather than testing of samples; sampling is discussed later). They are classified as a result of the tests, and sold or scrapped accordingly. For example, a resistor would be classified as high precision (typically ±2 percent), non-precision (±10 percent), or scrap. A microprocessor would be classified according to its maximum operating speed and other parameters.

Testing before assembly onto boards is very seldom necessary for discrete components, but might be appropriate for highly stressed power components such as power transistors, and complex ICs are sometimes tested before being soldered to boards. Other types of components, such as connectors, backplanes and unloaded circuit boards, must also be considered.

Another practical difference between electronic components and many non-electronic components, from the test and repair point of view, is that electronic components cannot be repaired or adjusted. It is not possible to repair a failed integrated circuit or resistor.

10.8.1 Integrated circuits

IC manufacturing processes are complex and relatively difficult to control, so the yields are relatively low. For example, the proportion of 'good' ICs on a silicon wafer is typically anywhere in the range 50 percent to 95 percent, depending on circuit complexity, process maturity, etc. Therefore it is absolutely necessary that every IC is tested as part of the manufacturing process, despite the high costs involved. (Some SSI/MSI circuits are nowadays not tested, or are given only simple tests.) Figure 10.7 shows the three categories of component that can be manufactured in a typical process. Most are 'good', and are produced to specification. These should not fail during the life of the equipment. Some are initially faulty and fail when first tested ('*dead on arrival*' or DOA), and are removed. They therefore do not cause equipment failures. However, a proportion might be faulty, but nevertheless pass the tests. The faults will be potential causes of failure at some future time. Typical faults of this type are weak wire bond connections, silicon, oxide and conductor imperfections, impurities, inclusions, and leaky packages. These components are sometimes called *freaks*.

Burn-in is a test in which the components are subjected to high temperature operation for a long period, to stimulate failure of faulty components by accelerating the stresses that will cause failure due to these faults, without damaging the good ones.

Test methods for integrated circuits (ICs) have been prescribed in standards, the earliest being US MIL-STD-883 (Test Methods and Procedures for Microelectronics) (Table 10.2). Similar methods are described in other national and

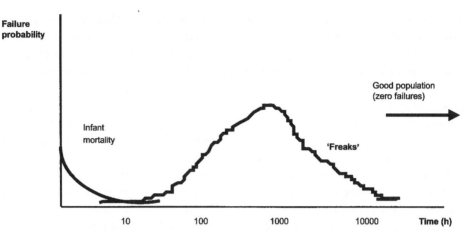

Figure 10.7 Quality categories of electronic components

Table 10.2 MIL-STD-883/BS9400 Microelectronic Device Screening Requirements*

Screen	Defects effective against	Level A	Level B	Level C
Pre-encapsulation visual inspection (30–200 × magnification)	Contamination, chip surface defects, wire bond positioning wire bond positioning	100%	100%	100%
Stabilisation bake	Bulk silicon defects, metallisation defects (stabilises electrical parameters)	100%	100%	100%
Temperature cycling	Package seal defects, weak bonds, cracked substrate	100%	100%	100%
Constant acceleration (20000 g)	Chip adhesion, cracked die, weak bonds	100%	100%	100%
Leak tests (hermetic packages)	Package seal	100%	100%	100%
Electrical parameter tests (pre-burn-in)	Functional failure	100%	100%	100%
Burn-in test (168 h, 125°C with applied AC voltage stress)	Surface and metallisation defects	100%	100%	N/A
Electrical test (post-burn-in)	Functional failure	100%	100%	N/A
X-ray	Particles, wire bond position	100%	N/A	N/A

* This table is not comprehensive, and the reader should refer to the appropriate standard to obtain full details of applicable tests.

international commercial standards (see Appendix 2). The standards set two (sometimes three) levels of quality. In the lower level (Level C), the components must be subjected to a range of inspections and functional tests. In the next higher level, which was generally specified for military and aircraft systems (Level B), they must also be subjected to a 100 percent high-temperature long-duration screening test (125°C for 168 hours). This temperature and duration were determined as considered appropriate for ceramic or metal packaged military grade components. The third level (Level A) is for components to be used in space applications. In fact these standard test plans are now very little used, and component manufacturers determine and apply their own tests.

Component manufacturers and users have developed variations of the standard screens and burn-in methods. The main changes are in relation to the burn-in duration, since 168 hours has been shown to be longer than necessary to remove the great majority of defectives (the only justification for 168 hours was merely that it is the number of hours in a week!). Also, more intensive electrical tests are sometimes applied, beyond the simple reverse-bias static tests specified. Dynamic tests, in which gates and conductors are exercised, and full functional tests with monitoring, are applied to memory devices, other VLSI/ULSI devices, and ASICs, when the level of maturity of the process or design and the criticality of the application justify the additional costs.

Plastic encapsulated components cannot generally be safely burned-in at 125°C, so lower temperatures are used. Also, in place of leak tests they are tested for moisture resistance, typically for 1000 hours at 85°C and 85 percent relative humidity (RH). This is not, however, a 100 percent screen, but a sample test to qualify the batch. A more severe test, using a non-saturating autoclave at 100°C and 100 percent relative humidity (RH), is also used, as the 85/85 test is not severe enough to show up problems with current encapsulating processes.

The decision on whether and how to test components as part of an equipment manufacturing operation must be based on the difference between the costs of test and the total downstream costs of not testing. A modern IC tester costs between $500 000 and $5 million, and this is increasing almost in proportion to the rise in complexity of the ICs being built and tested. This is a major reason why equipment and system manufacturers seldom test ICs. The specialised knowledge required also militates against component testing beyond the IC manufacturing stage. However, if the equipment production volumes are sufficiently high, or if there is sufficient uncertainty about the quality levels of the ICs being delivered, testing must be considered.

As stated earlier, the quality levels of most standard ICs are very high, so that they are very seldom tested by equipment manufacturers. However, the great improvements made since the early 1980s have not been consistent across all device types. Complex programmable logic devices (PLDs) and mixed signal devices have much poorer quality levels than simpler high-volume devices. Yield at wafer stage is quite poor for many devices and the only way to get from this point to shipping high-quality levels is by testing, but achieving 1 ppm defective, or even 100 ppm, starting from typical wafer yields requires levels of fault coverage not currently attainable with vector-based testing and test generation methods. I_{DDQ} testing (Chapter 8) offers a possible solution but device manufacturers have not yet embraced this fully due to the test time and the cost involved.

10.9 STATISTICAL PROCESS CONTROL AND ACCEPTANCE SAMPLING

The variation that is inherent in manufacturing processes can be monitored, controlled and improved using the technique of *statistical process control* (SPC). This is described in Chapter 12.

Statistical acceptance sampling is a method for setting criteria for the allowable proportion of items that can be defective in a sample, based upon the criteria set for the whole production run. It is a method that is incompatible with most modern manufacturing, and should be avoided except in special cases. It is also described in Chapter 12.

REFERENCES

1. Juran, J. M., 1974, *Quality Control Handbook*, McGraw-Hill.
2. Ishikawa, K., 1990, *Introduction to Quality Control*, Chapman & Hall.
3. Imai, M., 1997, *Gemba Kaizen*, McGraw-Hill.
4. Sydenham, P. H., Hancock, N. H. and Thorn, R., 1989, *Introduction to Measurement Science and Engineering*, John Wiley & Sons, Ltd.
5. Sydenham, P. H. and Hancock, N. H. Hancock, 1982–92, *Handbook of Measurement Science* (3 vols), John Wiley & Sons, Ltd.
6. Polak, T. A. and Pandy, T., 1999, *Engineering Measurements: Methods and Intrinsic Errors*, Mechanical Engineering Publications.
7. Hobbs, G., 2000, *Accelerated Reliability Engineering: HALT and HASS*, John Wiley & Sons, Ltd.
8. Davis, B., 1993, *The Economics of Automatic Testing* (2nd edn), McGraw-Hill.

11

Testing in Service

11.1 INTRODUCTION

As in the preceding phases of the life of an engineering product, the ideal test situation in service is zero testing. For many types of product this is achievable and practicable. We do not test our TVs, mobile phones or dishwashers to see if they are working correctly, but instead rely on the evidence of their condition as observed during use. On the other hand, we cannot tell whether the emissions from a car engine are within allowable limits without performing a test. Generally, we must consider testing items in service in the following circumstances.

- Their correct function cannot be assured otherwise. This includes requirements for accuracy, such as for measuring instruments. Such testing is part of preventive maintenance, i.e. maintenance to prevent failure.

- They are not working correctly, but we must perform diagnostic tests to find out why.

- They have been repaired or overhauled, and proof of correct function is necessary before they can be put back into service. This is part of corrective maintenance.

There are close connections between testing in service and in manufacture. The same or similar facilities (test equipment, software) are often used, and the design features that enhance testability in manufacture carry through into service. The test methods and technologies described in earlier chapters are therefore relevant, and in-service testing should be optimised during design and development.

In this chapter we will look at the technology and management aspects of in-service test. Management aspects will be further discussed in Chapter 14.

11.2 IN-SERVICE TEST ECONOMICS

Optimising the economics of in-service testing follows the same principles as are applicable to manufacturing test, as described in Chapter 10. We must take account of all of the practical aspects, such as the costs of test and repair facilities, the performance of test personnel and equipment, equipment utilisation, location and accessibility, the effects and costs of failures and the requirements for scheduled tests and calibration. The test economics problem can be modelled and the optimum decisions can be derived. Very often the decisions are clear-cut, for example:

• Domestic equipment: generally failed components are replaced, and are not repaired.

• Mobile telephone system: the expensive, critical equipment in ground stations is continually monitored. Failures are repaired by replacement of items such as circuit boards, which are then sent to a depot or factory for repair. However, failed individual handsets are exchanged and failed circuit boards are scrapped, since they are very difficult to diagnose and repair, but are relatively inexpensive to manufacture.

However, some situations are more complex or finely balanced, and require more detailed analysis. *Reliability-centred maintenance* (RCM) is one method that can be applied, and this is described later.

11.3   TEST SCHEDULES

When tests, or any other scheduled maintenance activity, are determined to be necessary, we must also determine the most suitable intervals between them. Maintenance schedules should be based upon the most appropriate time base. These can include:

• Road and rail vehicles: distance travelled

• Aircraft: hours flown, numbers of takeoff/landing cycles

• Electronic equipment: hours run, numbers of on/off cycles

• Fixed systems (radars, rail infrastructure, etc.): calendar time.

The most appropriate time base is the one that best accounts for the equipment's utilisation in terms of the causes of degradation (wear, fatigue, parameter change, etc.), and is measured. For example, we measure the distance travelled by our cars, and most degradation is related to this. On the other hand, there is no point in setting a calibration schedule based upon running time for a measuring instrument unless an automatic or manual record of its utilisation is maintained, so these are usually calendar-based.

The intervals must be based upon the rate at which degradation can occur and the consequences (cost, hazard, etc.) of failures that might be caused. The RCM process can be used to optimise maintenance intervals.

11.4　MECHANICAL AND SYSTEMS

The methods described in Chapter 7 apply to in-service testing of mechanical equipment and systems. There is a wide range of situations. For example:

- A pressure relief valve might be tested to ensure that it operates at the correct pressure.

- Jet engines are tested both on test stands after repair or overhaul, and on the aircraft after installation or in-situ repair, or for diagnostic checking.

- Large systems such as aircraft, rail vehicles, cars and trucks are tested regularly to ensure that important functions such as controls, brakes, etc. are functioning correctly. Tests such as these are part of the maintenance schedule, which includes other preventive maintenance tasks such as lubrication, cleaning, visual checks and replacements.

Monitoring methods are used to provide periodic or continuous indications of the condition of mechanical components and systems. These include:

- Non-destructive test (NDT) for detection of fatigue cracks

- Temperature and vibration monitors on bearings, gears, engines, etc.

- Oil analysis, to detect signs of wear or breakup in lubrication and hydraulic systems.

11.5　ELECTRONIC AND ELECTRICAL

Electronic components and assemblies generally do not degrade in service, so long as they are protected from environments such as corrosion. Electronic components and connections do not suffer from wear or fatigue, except as discussed in Chapter 2, so there is very seldom a pronounced 'wearout' phase in which failures become more likely or frequent. Therefore, apart from calibration for items like measuring instruments, scheduled tests are seldom appropriate.

Electronic units and systems can be tested in service using the same test equipment as is used for development and manufacture, as described in Chapter 8. There are four basic approaches that can be applied:

1. Use manual test instruments, such as oscilloscopes, RF testers, etc., as used in development testing.

2. Use a '*hot rig*', which is a complete working system into which units to be tested can be substituted, then tested as part of the system test.

3. Use production test equipment such as ICT and functional testers. This usually requires that items to be diagnosed or tested after repair must be

routed through the manufacturing test process, since it is not always economic
to provide such expensive test equipment separately for maintenance support.

4. Use test equipment designed specifically for maintenance support.

In fact there will often be overlaps between these approaches, depending on the
type of system and its application.

Specialist test equipment is widely used for in-service testing. These are often
compact and ruggedised versions of laboratory or factory test equipment, as
shown in Figure 11.1. Since a large proportion of the failures on many types of
electronic systems in service are caused by cable and connector faults (primarily
opens and intermittents), cable testers are commonly used to locate them. The
time domain reflectometer (TDR) is a useful instrument that indicates the
distance to a fault by transmitting a pulse and measuring the time to return
from the high impedance at the break. Insulation resistance tests are performed
to check the integrity of cable insulation. Specialised mobile testers are available
for testing telecommunications equipment, etc. Such testers are used when it is
not practicable or economic to take the equipment to the test facility, so the tester
must be taken to the equipment.

For fixed installations such as telecommunications systems, railway signalling,
process plant, etc., much of the testing must be performed on location, so

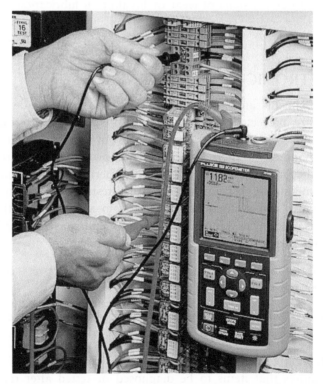

Figure 11.1   Service test equipment (courtesy Fluke)

portable instruments which combine diagnostic and functional test capabilities are used. These are often specially developed for the system being tested. The same applies to some testing on systems such as aircraft. In most cases repairs on such systems are performed by replacement of circuit cards or of larger assemblies such as boxes ('*line replaceable units*' (LRU) is a common expression for such items), and these are returned to a depot or the factory for further diagnosis and repair.

It was quite common for some large operators, particularly the military, to develop large, multi-purpose test systems which could be used for depot testing of a wide range of different electronic units. The US Navy's VAST (Versatile Avionic Shop Test) for use on aircraft carriers and the system developed for the European Tornado aircraft were examples. However, it was found that the expected benefits of these systems were not achieved, due to the difficulty of managing the interfaces between the designers of the test system and all of the separate companies developing the units to be tested. Test systems such as these must be continuously updated (hardware, interface adaptors, software, operator instructions, etc.) in order to cater for changes in the units to be tested, and this proved to be very difficult and expensive. Also, the attempt to replace the need for skilled technicians with general-purpose ATE backfired: the ATE could never learn from experience, so every test on a particular item always followed the same sequence. Experience showed that it is far more effective for the designers of electronic equipment to also design and provide the in-service test equipment, optimised for that item rather than compromised over a wide range of similar ones.

11.5.1   Built-in test

Many electronic systems include *built-in test* (BIT) (also referred to as *built-in self-test* (BIST)) for in-service testing. For example, computers usually run tests on memory and other functions on start-up. BIT can also be designed to operate continuously or on demand. BIT must be specified, designed and developed concurrently with other aspects of the system. Guidelines for the application of BIT are:

- BIT should not be applied to functions whose correct or incorrect operation would be apparent to the operator. For example, if the cursor on a radar screen shows whether the scanner is operating, there is no benefit in having BIT monitor this function.

- BIT should be kept simple. BIT can generate false alarms, and the more functions that are monitored the greater will this problem be. In particular, the use of additional hardware, such as sensors, wiring, etc., should be minimised.

- As far as practicable BIT should be implemented in software. Software adds no weight, complexity or power requirements, and it does not fail in the ways that additional hardware can.

11.5.2 'No fault found'

A large proportion of the reported failures of many electronic systems are not confirmed on later test. These occurrences are called *no fault found* (NFF) or *re-test OK* (RTOK) faults. There are several causes of these, including:

- Intermittent failures, such as components that fail under certain conditions (temperature, etc.), intermittent opens on conductor tracks or solder connections, etc.

- Tolerance effects, which can cause a unit to operate correctly in one system or environment but not in another

- Connector troubles in which the failure seems to be cleared by replacing a unit, when in fact it was caused by the connector that is disturbed to replace it

- BIT systems which falsely indicate failures that have not actually occurred

- Failures that have not been correctly diagnosed and repaired, so that the symptoms recur

- Inconsistent test criteria between the in-service test and the test applied during diagnosis elsewhere, such as the repair depot

- Human error or inexperience

- In some systems, the diagnosis of which item (card, box) has failed might be ambiguous, so more than one is replaced even though only one has failed. Sometimes it is quicker and easier for the technician to replace multiple items, rather than trying to find out which has failed. In these cases multiple units are sent for diagnosis and repair, resulting in a proportion being classified NFF. (Returning multiple units can sometimes be justified economically. For example, it might be appropriate to spend the minimum time diagnosing the cause of a problem on a system such as an aircraft or oil rig, in order to return the system to operation as soon as possible.)

The proportion of reported failures that are caused in these ways can be very high, often over 50 percent and sometimes up to 80 percent. This can generate high costs, in relation to warranty, spares, support, test facilities, etc. NFF rates can be minimised only by effective management of the design in relation to in-service test, and of the diagnosis and repair operations. Stress screening of repaired items can also reduce the proportion of failures that are not correctly diagnosed and repaired, as described later.

11.6 SOFTWARE

As discussed in Chapter 9 software does not fail in the ways that hardware can, so it is not appropriate to test programs that are being used. If it is found to be necessary to change a program for any reason (change in system requirements, correction of a software error), this is really redesign of the program, not repair. So long as the change is made to all copies in use, they will all work identically, and will continue to do so.

11.7   RELIABILITY CENTRED MAINTENANCE

Reliability centred maintenance (RCM) is a technique that has been developed to optimise preventive maintenance. The objective is to identify those tasks (inspection, test, replacement, etc.) that should be performed to generate the optimum balance between the likelihood of failure, in relation to the causes, patterns and effects of failures and the costs of prevention. RCM is widely applied in several industry sectors, particularly in transport, process industries and the military. It is described in Reference 1.

The RCM approach is based upon a sequence of questions and decisions in relation to every failure that is to be prevented or minimised. The questions relate to:

1. The pattern of the failure, in relation to time, operating cycles, etc. If the pattern presents an increasing risk, such as caused by fatigue, wear, corrosion, etc., then scheduled maintenance action (check, repair, replace, etc. as appropriate) is considered.

2. The effect (including cost) of the failure. If the effect is slight or inexpensive (e.g. domestic light bulb), the item could be replaced when it fails. If the effect is critical or expensive (e.g. jet engine turbine disc) the item should be replaced before it fails. This information should have been generated as part of the failure modes and effects analysis (FMEA) performed on the design (Chapter 5).

3. The indications that are available of incipient failure, such as cracks, noise, vibration, etc. This information is used to determine the kinds of checks or tests that should be applied.

4. The amount of time (or distance, operating cycles, etc.) available between the first indication of potential failure and actual failure. This determines how often the item should be checked or tested.

The RCM process is shown in Figure 11.2.

Testing, as a preventive maintenance task, is one of the activities addressed by RCM. The principles described earlier apply: scheduled tests should be performed only if the RCM questions show that they are necessary. There is no value in performing scheduled tests of the following:

- Nearly all electronic equipment, since most failure modes are not caused by wearout mechanisms (Chapters 2 and 3). Also, failures nearly always occur

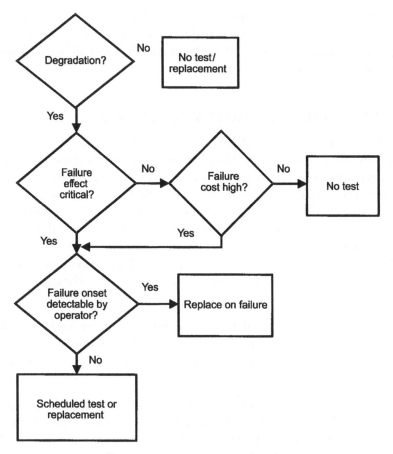

Figure 11.2   RCM process flowchart

without warning or prior indication. In most cases failures would be noticed by the operator or reported by the BIT (for example, TV sets, PCs, printers, etc.). However, for items for which incorrect or inaccurate operation cannot be observed (emergency radios, measuring instruments, etc.), tests and calibration should be considered.

- Any item whose condition can be observed by less expensive means, such as inspection.

11.8   STRESS SCREENING OF REPAIRED ITEMS

The stress screening methods described in Chapter 10 can sometimes be applied effectively to items that have been repaired. The objective is the same: to show up weaknesses that are not apparent during tests under non-stress conditions. Stress screening is particularly effective in finding the causes of intermittent failures, or other causes of 'NFF'. It can be used to assist diagnosis, by showing up failures that are not apparent otherwise, and to validate that repairs have been performed correctly. Of course it is important to ensure that the stresses do not cause damage to good items.

Ideally, the HASS method should be applied, which in turn implies that the design has been subjected to HALT so that the stress regime is optimised for the product.

The use of stress screening, and particularly HASS, as part of the repair process has been found to be very effective in improving reliability and reducing costs. It is one of the many important benefits that can be achieved by ensuring that testing is managed systematically and consistently during development, manufacture and service.

11.9 CALIBRATION

Calibration is the regular check or test of equipment used for measuring physical parameters, by making comparisons against standard sources. Calibration is applied to basic measurement tools such as micrometers, gauges, weights and torque wrenches, and to transducers and instruments that measure parameters such as flow rate, electrical potential, current and resistance, frequency, etc. Therefore calibration may involve simple comparisons (weight, timekeeping, etc.) or more complex testing (radio frequency, engine torque, etc.).

Whether an item needs to be calibrated or not depends primarily upon its application, and also upon whether or not inaccuracy would be apparent during normal use. Any instrument used in manufacturing should be calibrated, to ensure that correct measurements are being made. Calibration is often a legal requirement, for example of measuring systems in food or pharmaceutical production and packaging, retailing, safety-critical processes, etc.

All calibrations must be 'traceable' to an approved international or national standard. The International Standard metre, second, ohm, etc. are all defined by the International Standards Organisation (ISO) in Geneva. National 'primary reference standards' are based upon these, and are used as the basis for calibration in the relevant countries. *Traceability* means, in this context, that the accuracy of the measurement against which the item has been calibrated has in turn been calibrated against a 'higher' standard, and so on up to the relevant national or international reference standard. For example, slip gauges are used in manufacturing industry for calibrating measuring instruments like micrometers and co-ordinate measuring machines. In turn the slip gauges must be calibrated against a master set, which might be held locally or at a calibration centre, and these against the national reference. Calibration centres are 'accredited' by the relevant national organisation. The same approach applies to instruments such as weighing machines, flowmeters, electronic instruments, etc.

The accuracy required in calibration varies depending on the level. Typically, if a measurement uncertainty of not more than 1 in 10^4 might be acceptable for use, 1 in 10^5 would be suitable for local calibration, 1 in 10^6 for calibration at an accredited facility, and the national reference would have an uncertainty in the range 1 in 10^7 to 1 in 10^9. The actual values would depend upon the parameter. The international standard relating to accuracy of measurement and calibration is ISO5725.

Calibration may not be necessary for some instruments used in development areas, when accuracy is not critical. Also, regular calibration is not applied to items sold and used in non-critical applications such as private use.

Some electronic instruments include built-in self-calibrating facilities. These provide assurance that the instruments are accurate to the extent that the built-in capability provides. However, they do not replace the need for calibration against higher references. Electronic instruments can, in some cases, be calibrated remotely, using Internet-based communication of results and adjustments.

The main national organisations involved in calibration are listed in Appendix 2. There is an increasing trend towards international and transnational operations for test and calibration.

When calibration is required, it is essential that it is managed in a disciplined way. The basic requirements are:

- There must be a calibration schedule for every item, and if an item has not been calibrated by the end of the permitted period it must be taken out of use, or used only in applications for which calibration is not necessary.

- There must be a system for marking or labelling items to show their calibration status. Labels with green backgrounds are often used to indicate 'in calibration', and red to indicate 'out of calibration'.

- There must be a system for managing and recording the calibration requirements and status of all relevant instruments.

Calibration management systems are available, which provide automatic alerts and maintain records.

The international standard for quality management systems (ISO9000), described in Chapter 13, includes requirements for calibration. The international standard for the accreditation of measurement and test laboratories is ISO17025.

Calibration requirements and systems are described in Reference 2.

REFERENCES

1. Moubray, J., 1991, *Reliability Centred Maintenance*, Butterworth-Heinemann.
2. Morris, A. S., 1997, *Measurement and Calibration Requirements for Quality Assurance to ISO9000*, John Wiley & Sons, Ltd.

12

Data Collection and Analysis

12.1 INTRODUCTION

Throughout the test programme (development, manufacture, in-service) it is essential that data on tests performed and the results is collected and analysed. Without such an effort, there can be little or no improvement to the design or to the processes. The whole effort should be managed so that tests are planned, data is collected and analysed, improvement or corrective opportunities are identified, and action is taken. Every step is essential, since missing any one will break the improvement chain. Also, the reporting, particularly on failures or other serious problems during development and manufacture, should be total.

Test and failure data can be collected and analysed manually or automatically. *Data acquisition* (DAQ) hardware (sensors, signal conditioning and recording equipment) and software are available. The software typically can perform the analysis functions described later in this chapter, as well as data visualisation such as waterfall and rainflow analysis (Chapter 2), etc. Data acquisition and analysis features are built in to many modern test systems. The mathematical analysis software described in Chapter 5 also provides such functions.

Test results, particularly during development, nearly always include data on failures. Every failure that occurs (or that seems likely to occur) should be considered as the first warning of future problems. Every failure should, in principle, be prevented from occurring again. This philosophy might seem to be drastic and impracticable, but the long-term effect is to force continuous improvement and to greatly reduce the total number, and therefore cost, of failures downstream in the project cycle. It is not uncommon for failures on a test to be ignored on the grounds that they are considered not to be relevant to the test, or because other project priorities discourage reporting and analysis. This short-term thinking is almost always counter-productive, particularly when it occurs during development testing. The management of failures is sometimes referred to as a *failure reporting, analysis and corrective action system* (FRACAS).

12.2   FAILURE REPORTING, ANALYSIS AND CORRECTIVE ACTION SYSTEM (FRACAS)

The management system for dealing with failures should be established as a strict, written procedure, that permits waivers from the 100 percent discipline only under defined exceptional conditions. The FRACAS is normally a mandatory feature of quality system audit and registration in accordance with standards such as ISO9000. The system should include:

- Written procedures that describe responsibilities for data collection, retention, analysis, review and action
- Failure reporting forms
- Analysis methods
- A failure review and management process.

In this chapter we will look at each of these in turn.

12.2.1   Procedure for FRACAS

The FRACAS procedure is one of the few for which a mandatory, disciplined approach is justified (others relevant to testing are design records and calibration). Without it, there will inevitably be pressure to ignore failures or other problems, or to take short-cut approaches, in order to keep to schedules and costs on parts of the programme, to the detriment of longer-term benefits. Therefore the procedure must force rapid and full investigations, not merely token reporting. Large development programmes can generate many failure reports, which, if allowed to collect and are then reviewed later, say once a month, can result in meetings that have to deal with more than can be handled effectively. Furthermore, essential supporting information, such as the exact test conditions or the state of the failed item, are lost. In principle, all failures should be investigated immediately after they occur, with the people and the items involved present. We must identify the 'smoking gun'. Failed hardware should not be repaired before it is thoroughly investigated, or essential evidence might be lost.

The person responsible for ensuring that the procedure is applied should be the project manager, or the test manager on his behalf. Responsibilities of supporting people (failure investigation labs, reliability analysis, etc.) should all be defined.

When failed hardware must be investigated using special facilities, such as microscopy, X-ray, test instruments, etc., or sent to suppliers for investigation, it is important that the same priorities and disciplines are applied. Every failure investigation must be connected to the original failure, and the investigation results must be reported back to the failure review. A priority system should be imposed on all supporting investigations, related to the importance of the original failure.

Failed hardware that is not repaired or investigated should not be scrapped, but should be identified and retained for a suitable period in a bonded store. This provides an inexpensive way of ensuring that failed items are available for later investigation if necessary. It is particularly effective when applied during manufacture, and sometimes also in service.

12.2.2 Failure reporting forms

Failure reporting forms should be designed to make reporting comprehensive and consistent. They should be as simple as possible, and should avoid the need for complex or subjective coding for aspects such as causes of failure. A little information quickly is nearly always more effective than a lot of information later.

A suggested failure report form for use during development is shown in Figure 12.1.

Similar forms can be used for manufacturing and in-service. However, the approach used will obviously vary depending upon aspects such as manufacturing rates and quantities and service conditions. For low-rate manufacturing followed by service with close support from the manufacturer, the system used during development can be extended to cover the later phases. For high-rate manufacture, with emphasis on automatic test and measurement, the failure data system must be tailored to accept and, where appropriate, analyse the outputs from the manufacturing control system. Some ATE, machining and metrology systems include these facilities.

The requirements for failure data reporting and analysis in service are often determined by the customer, by the user or by regulation. Civil aviation, military, and other large organisations impose particular disciplines and systems, and OEMs such as car and other equipment manufacturers often dictate the requirements for lower-tier suppliers. In other situations failure data collection from service use might be difficult or impossible, for example for consumer equipment outside the warranty.

Having common systems through all product phases can be useful in harmonising aspects such as coding and in maintaining discipline, and in showing connections between development, manufacturing and in-service failures.

Sometimes a two-tier approach can be effective, particularly for equipment in service when full information on every failure might not be practicable. The first tier then should collect the basic information on all failures, and the second tier is then applied selectively to failures that are determined to be more important, on the basis of cost, risk, frequency, etc. The two-tier approach can be an effective way to optimise the efforts of service engineers. For first-tier failures they need report only the fact of the failure and other quick information, but for second tier they would be required to take defined actions, such as call for detailed instructions, return failed hardware for investigation, etc.

Failure Reporting Form

Report No.				
System		Serial No.	Date / /	Time :
Location/customer			Run time	h: min
Test/operation	Tick:	1 2 3 4 Schedule 1 Schedule 2 Schedule 3 Calibration Unscheduled Other		

Describe
(Conditions, special features)

Result/failure

Repair action

Parts replaced

Item	1	2	3	4	5
Part No.					
Serial No.					
Mod strike					

Time started :	Time finished :	Time worked h: min
Work done by:	Checked by:	

Analysis

Previous occurrences (Report Nos.)	Trend	Cause/s	Comments

Corrective action

Recommended

Agreed

Tested

Accepted

Approved	Date / /

Figure 12.1 FRACAS failure reporting form

12.2.3 Failure data analysis: the 'seven tools'

All failure data must be retained and made available for analysis. A database should be designed, and all relevant information should be included. The best way to achieve this is to use forms on a PC screen, supported by database software that automatically updates the database.

There are several ways in which test and failure data can be analysed. The most basic, and often the simplest, is to look at what has happened on an individual basis: if a transistor overheats or a spring breaks the reason is often apparent without the need to collect and analyse data on multiple events. The principle of *Ockham's Razor* should be applied: the best method or explanation is the simplest one that solves the problem.

The *'seven tools of quality'* are simple methods for analysing data, that were identified as being most appropriate for applying to manufacturing quality improvement in the 'Quality Circles' movement in Japan. They can be applied to other situations, including development and in-service, as appropriate. (They can also be applied to non-engineering situations.) They are effective because they are simple, and can therefore be taught to and applied by people who do not have specialist knowledge or skills in statistical methods.

The seven tools are:

1. Brainstorm

2. Data collection

3. Data analysis methods, including measles charts, trend charts and regression analysis

4. Pareto chart

5. Histogram

6. Cause-and-effect (or Ishikawa) diagram

7. Statistical process control (SPC) chart.

Brainstorm

A *brainstorm* is a structured meeting to discuss a subject or a problem, in which the team concentrates on the topic in a disciplined way to determine a solution. One person always leads the discussion and ensures that it is planned and conducted in a defined sequence. First the problem is presented and agreed upon. Then participants are encouraged to suggest possible causes, without any further discussion or criticism at this stage. This is essential to ensure that all possible ideas are put forward and noted. Only when this aspect has been completed, and no further ideas are presented, the ideas are then discussed in turn, in relation to their likely importance in relation to the problem, and then in relation to suggestions, recommendations and action, again in order.

Data collection

There are many ways in which data can be collected. Figure 12.2 shows a simple tally chart that can be used, for example, for recording the number of defects on items that have passed through a process (painting, assembly, soldering, etc.).

Data analysis

Figure 12.3 shows the most common analysis methods used by Quality Circles teams.

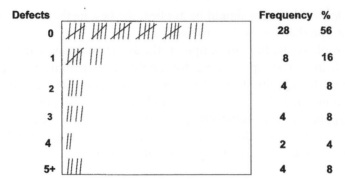

Figure 12.2   Checksheet (tally chart)

- *Measles chart.*   The *measles* chart is a picture of the item (machined part, loaded circuit board, painted assembly, etc.), on which are marked the locations of problems. Different kinds or colours of 'spots' can be used to indicate different problems, times of day, etc. The chart shows the patterns of problems that occur, to provide clues to the possible causes.

- *Trend charts.*   Trend charts can be used to monitor the behaviour of items or processes over time, cycles, etc., and to compare the behaviour. For example, if two machines are producing the same component, then the trend charts can show if they are both performing similarly or differently, and when trends occur.

- *Regression charts.*   A regression chart is a graph of one variable (measured quantity) against another. If the points follow a definite trend, say a straight line, then the two variables are probably correlated: one influences the other.

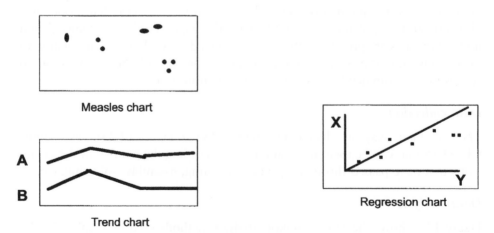

Figure 12.3   Data analysis methods

Statistical regression analysis can be used to determine the correlation coefficient, but in most cases simple observation is sufficient to enable practical judgements to be made.

Pareto Chart

A *Pareto* chart (Figure 12.4) (named after Vilfredo Pareto, the Italian economist who studied income distributions) is a simple method for ranking data so that the most important aspects are highlighted. For example, the columns in Figure 12.4 could relate to failed components within a system or any other failure causes. The vertical scale can represent quantity, proportion or costs. Pareto charts can be extended to analyse lower levels of problems; for example if a particular component causes 60 percent of failures, then the contributing reasons for this can be analysed on a separate chart, or within the columns on the main chart.

Histogram

A *histogram* (Figure 12.5) is a chart of a variable quantity against frequency of occurrence. Histograms show the shapes of distributions and the major properties (average, spread). Again, statistical methods can be used to analyse the data more rigorously, but, as explained in Chapter 4, simple observation is usually sufficient.

Cause-and-effect diagram

The *cause-and-effect diagram* (also called the '*fishbone*' diagram, or the *Ishikawa* diagram after its inventor) (Figure 12.6) provides a method for structuring and recording the brainstorm discussions and conclusions. The main problem is identified, then as the possible contributory factors are identified they are written against the 'fishbones'. The diagram can be set out initially to direct the thinking

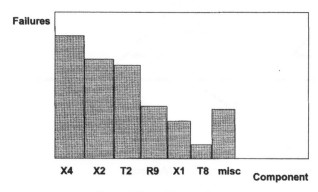

Figure 12.4   Pareto chart

Data collection and analysis

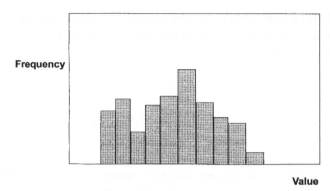

Figure 12.5 Histogram

along particular lines, such as the 'four Ms' (men, machines, methods, materials). First, ideas are collected and recorded. Then they are revisited in turn, with the emphasis on prioritising and on determining solutions or recommendations for action. The other six tools can be used as appropriate for further analysis.

Statistical process control chart

Statistical process control (SPC) charts are used to plot the variation of measured parameters, so that processes can be monitored and improved. Figure 12.7 shows an example. The charts comprise two separate plots that are created concurrently as the process proceeds. The *means chart* (or $\bar{X}$ *chart*) is used to record, for each batch, the average value of the measured parameter. The *range chart* (or $\bar{R}$ *chart*) is used to record the range between the highest and lowest values within the batch. Therefore the $\bar{X}$ chart indicates the accuracy of the process, and the $\bar{R}$ chart indicates its precision.

The *control limits* on the charts are derived from analysis of the variation of production trials, to determine the *process capability*. This analysis is based upon

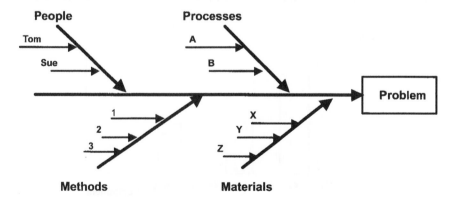

Figure 12.6 Cause and effect diagram

the statistical ideas described in Chapter 4, particularly use of the normal distribution, so the control limits are determined from the estimated standard deviation of the trial sample. As explained therein, however, such mathematical finesse can be misleading. W.A. Shewhart, the inventor of SPC charts, emphasised that they should be used primarily for driving process improvements, by identifying special causes of variation, as described in Chapter 4. For example, two likely assignable causes are: a regular sawtooth pattern, which might be caused by a tool change, shift change, temperature controller variation, etc., and a consistent high value, which might be due to a measurement error, etc.

The seven tools are described in more detail in several books. References 1–4 are good examples.

12.3 ACCEPTANCE SAMPLING

Statistical acceptance sampling is a method that was developed to determine whether a production batch contains not more than a predetermined proportion of defective items, called the *Acceptable Quality Level* (AQL). The method is based upon testing or inspecting a randomly selected sample of the production batch, and then comparing the number of defectives with the statistical criteria. The criteria are determined using the binomial distribution statistics, as described in Chapter 4, Reference 1 and in all books on statistical quality control, and in standards such as US MIL-STD-105 and British Standard 6001. For example, for an AQL of 0.1 percent and a sample size of 500, the batch would be rejected if there were two or more defective, and accepted if there were one or zero defective. A more stringent sampling procedure was introduced later to deal

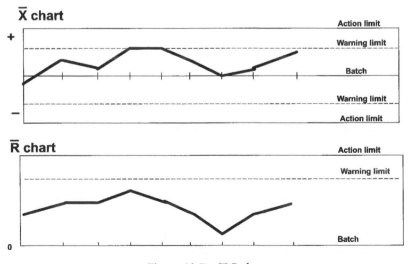

Figure 12.7 SPC charts

with items with much lower expected proportions defective, such as electronic components: this is the *lot tolerance percent defective* (LTPD) method.

All statistical sampling methods rely on the inspection or testing of samples that are drawn at random from the manufacturing batches, then using the mathematics of probability theory to make assertions about the quality of the batches. The sampling plans are based upon the idea of balancing the cost of test or inspection against minimising the probability of the batch being accepted with an actual defective proportion higher than the AQL or LTPD. Since the samples are random, it is always possible that the results might be 'lucky' or 'unlucky', so there are risks of wrong decisions being made. These are called the *supplier's risk* and the *customer's risk*, and they are quantified as percentage probabilities by the methods.

However, optimising the cost of test or inspection is not an appropriate objective. The logically correct objective in any test and inspection situation is to *minimise the total cost of manufacture and support*. When analysed from this viewpoint, the only correct decision is to perform either 100 percent or zero test/inspection. There is no theoretical sample size between these extremes that will satisfy the criterion of total cost minimisation. In addition, most modern manufacturing processes, particularly at the level of components, generate such small proportions defective (typically a few per million) that the standard statistical sampling methods such as AQL and LTPD cannot discriminate whether or not a batch is 'acceptable'. For these reasons statistical sampling is very little used nowadays.

The fundamental illogic of statistical acceptance sampling was first explained by Deming (Reference 5). If the cost of test or inspection of one item is k_1, the cost of a later failure caused by not inspecting or testing is k_2, and the average proportion defective is p, then if p is less than k_1/k_2 the correct (lowest total cost) strategy is not to test any. If p is greater than k_1/k_2 the correct strategy is to test all. This explanation represents the simplest case, but the principle is applicable generally: there is no alternative theoretically optimum sample size to test or inspect. The logic holds for inspection or test at any stage, whether of components entering a factory or of assembled products leaving.

For example, if an item costs $50 to test at the end of production, and the average cost of failure in service is $1000 (warranty, repair, spares, reputation), then $k_1/k_2 = 0.05$. So long as we can be confident that the production processes can ensure that fewer than 5 percent will have defects that will cause failures in service, then the lowest cost policy is not to test or inspect any.

The logic of 0 percent or 100 percent test or inspection is correct in stable conditions, that is, p, k_1 and k_2 are known and are relatively constant. Of course this is often not the case, but so long as we know that p is either much larger or much smaller than k_1/k_2 we can still base our test and inspection decisions on this criterion. If we are not sure, and particularly if the expected value of p can approach k_1/k_2, we should test/inspect 100 percent.

There are some situations in which sampling is appropriate. In any production operation where the value of *p* is lower than the breakeven point but is uncertain, or might vary so that it approaches or exceeds this, then by testing or inspecting samples we might be able to detect such deviations and take corrective action or switch to 100 percent inspection/test. It is important to note, however, that there can be no calculated optimum statistical sampling plan, since we do not know whether or by how much *p* changes. The amount and frequency of sampling can be determined only by practical thinking in relation to the processes, costs and risks involved. For example, if the production line in the example above produces items that are on average only 0.01 percent defective, at a rate of 1000/week, we might decide to test 10/week (HASA, or other appropriate test) as an audit, because 10 items can be fitted into a test chamber with minimum interruption to production and delivery.

Items that operate or are used only once, such as rivets or locking fasteners, airbag deployment systems, and pressure bursting discs, can be tested only on a sample basis, since 100 percent testing of production items is obviously not feasible. The optimum sample plan is still not statistically calculable, however, since the proportions defective are usually much lower than can be detected by any sample, and it will nearly always be highly uncertain and variable. Inspection and measurement (100 percent or sample) are important to ensure uniformity of the production items.

12.4   PROBABILITY AND HAZARD PLOTTING

Probability plotting is a graphical technique for the statistical analysis of data, which uses specially constructed graph papers to determine the types and parameters of the distributions that fit the data. All probability papers use specially transformed scales, so that if the plotted data lies on a straight line on the graph it is an indication that the data fits the distribution for which the paper is designed. Graph papers are available for all the common distributions, including normal, Weibull, exponential, etc. Hazard plotting is a related method in which the cumulative hazard rate is used instead of the cumulative proportion.

Figure 12.8 shows an example of a probability plot, using Weibull graph paper, of the times to failure of components on test. The times to the first, second, etc. failures are plotted against the cumulative proportion that has failed at each time. The location and slope of the 'best fit' line through the data points allow the distribution parameters (in this case the mean life μ and the slope parameter β) to be estimated graphically.

It is important that probability and hazard plotting methods are used only to analyse data from situations in which the separate events (failures, etc.) occur independently of one another, and that the times to the events (first, second, etc.) are identically distributed. Statisticians call such data *independently and identi-*

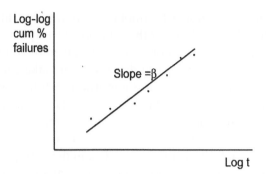

Figure 12.8   Weibull probability plot

cally distributed (IID). If the data is not IID the probability plotting method can generate misleading results. Therefore the method is suitable for analysing tests results on items such as components that can fail only once (fasteners, electronic components, etc.). It should not be used for analysing data on items that can fail more than once, such as repairable systems, for which the method described in the next section is appropriate.

There are several further aspects of probability and hazard plotting that should be understood and applied in practical situations. References 6 and 7 describe the methods in detail, and Reference 8 describes applications to accelerated test.

12.5   TIME SERIES ANALYSIS

Time series analysis (TSA) is a statistical method which is appropriate for analysing any data that arises as events over time, such as failures of repairable machines or systems. It is very simple, and it can be performed graphically. Figure 12.9 is a time series chart of failures of a product over time while in service. The three symbols denote different causes of failure, and the vertical lines indicate the equipment's scheduled tests. Four equipments are shown, and the bottom line shows the total pattern of failures.

The chart shows clearly that:

- Failures indicated by the symbol ▲ tend to arise most often just after the scheduled tests, and in clusters. Therefore they are probably caused by the test itself, and the clustering is a clue that there might be problems with diagnosis or repair.

- Failures of type ▼ tend to occur late in the scheduled test cycle. Therefore they are probably caused by a wearout or drift mechanism.

- Failures of type ■ do not appear to have any particular pattern.

Time series analysis is a very effective and flexible method for analysing test or in-service data. The horizontal lines can show individual equipments, fleets, customers, locations, modification states, etc. The 'time' scale could indicate running time, calendar time, operating cycles, etc.

Time series analysis is one of a number of methods collectively called *exploratory data analysis*. References 6 and 9 describe the methods.

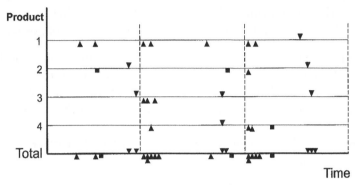

Figure 12.9 Time series analysis chart

12.6 SOFTWARE FOR DATA COLLECTION AND ANALYSIS

Many modern inspection and test systems (co-ordinate measuring machines, automatic test equipment, etc.) include capabilities for data collection and analysis, or for submitting data to analysis software.

All of the analysis methods described above can be performed using suitable statistical software. Spreadsheets can also be used for some analyses.

12.7 RELIABILITY DEMONSTRATION AND GROWTH MEASUREMENT

Statistical methods have been developed and applied for demonstrating the reliability of a product and for measuring the growth of reliability, during tests or in service. They are based upon calculating the mean time between failures (MTBF) and comparing it to the product requirements.

12.7.1 Reliability demonstration

Reliability demonstration testing (RDT) is based upon the method of *probability ratio sequential testing* (PRST). The equipment being evaluated is operated for a period of time expressed as multiples of the specified MTBF, and as failures occur they are plotted against the operating time. Testing is continued until the 'staircase' plot of failures against time crosses a decision line. The decision lines are constructed from the test criteria that have been decided beforehand. These are:

- The upper test MTBF, θ_0, the level considered to be 'acceptable'
- The lower test MTBF, θ_1, the agreed minimum value to be demonstrated
- The design ratio, $d = \theta_0/\theta_1$
- The decision risks, expressed as percentages. The 'supplier's risk', α, is the probability that equipment with an actual MTBF that is higher than required to pass will by chance fail. The 'customer's risk', β, is the opposite.

The method is the basis for US MIL-HDBK-781 and other standards, as described in Appendix 2. Figure 12.10 shows an example of a PRST plot.

The tests are required to be performed using operating and environmental conditions that represent expected in-service conditions, such as shown in Figure 12.11.

This method *should not* be used, particularly during development testing, for the following reasons:

- It implies that all failures are of equal importance, so that simple summation can provide the main criterion.

- It implies that all failures occur at constant average rates in time, in particular that there are no significant wearout failure modes which could cause increasing numbers of failures later.

- It ignores action taken to correct causes of failures that have occurred.

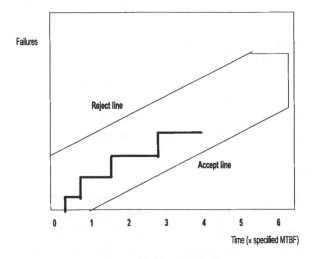

Figure 12.10 PRST plot

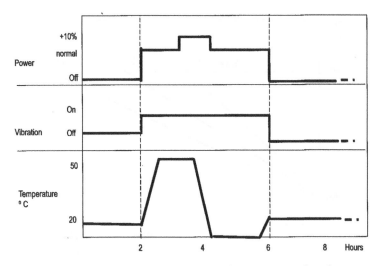

Figure 12.11   MIL-STD-781 typical environmental cycles

- It ignores failures that have *not* occurred, but that might occur in future.

- By being based upon *simulated* instead of *accelerated* stress conditions, it does not effectively or economically stimulate failures to identify opportunities for improvement, as described in Chapter 5.

- It generates an attitude that discourages the occurrence of failures, since the objective is to pass the test rather than to generate improvement opportunities. The parties concerned argue about aspects such as whether or not particular failures should be 'counted' and statistical interpretations, instead of concentrating on the engineering.

The PRST method is described and discussed in more detail in Reference 6. It is unfortunate that it is sometimes still taught and applied, and that standards exist (see Chapter 13 and Appendix 2) that call for its use.

12.7.2   Reliability growth monitoring

Reliability growth monitoring (RGM) is the term used to cover a range of statistical methods that have been developed for comparing the MTBF measured during tests or in service with targets or requirements. The best known is the *Duane method*. The cumulative MTBF, measured at intervals, is plotted and compared with the target or required value, using logarithmic scales. The slope gives an indication of the rate of MTBF growth, and extrapolations indicate when the target or requirement might be achieved. Figure 12.12 shows an example.

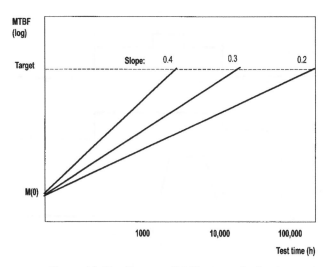

Figure 12.12   Duane reliability growth chart

These methods are subject to most of the same criticisms that apply to reliability demonstration tests. *RGM should never be used for monitoring reliability growth during development testing*, since the objective in this phase should be to generate failures as quickly as possible so that real reliability growth can be achieved by identifying and implementing improvements. It can be used as part of an in-service monitoring process, provided that it is supplemented by other methods that can help to identify failure causes and priorities, such as Pareto plots, time series analysis, etc.

The methods are described and discussed in more detail in Reference 6.

12.8   GENERAL COMMENTS ON DATA ANALYSIS

As with any other analytical approach, *Ockham's Razor* applies: the simplest method that provides the correct answer is the best. It is also important to heed the words of Kendall and Stuart, authors of the classic textbook *The Advanced Theory of Statistics*: '*A statistical relationship, however strong and however suggestive, can never establish a causal connection. Our ideas on causation must come from outside statistics, ultimately from some theory.*' We must always seek the true cause (physical, chemical, human, etc.) for problems and variation. As explained in Chapter 4, variation in engineering can be extremely 'noisy' and uncertain, so the statistical methods must be used to provide clues. They cannot provide explanations.

It is also worth remembering that truly random, independent events always show patterns of occurrence. Lotteries are random, but the different numbers do not come up at the same frequencies over any finite number of runs, and aircraft crashes often seem to occur in clusters. Therefore clustering or trending on a chart is not by itself evidence of a significant pattern.

Statistical significance tests, as described in Chapter 4, can be performed to indicate the likelihood of a set of observations occurring by chance. For example, a cannon-launched projectile showed an overall success rate of 90 percent on firing trials, in which X were fired. Further investigation showed that those fired over longer ranges had a lower success rate than those fired over shorter ranges. Was this significant? Statistical significance testing showed that it was: longer ranges seemed to cause lower reliability. Since the longer-range firings meant that a larger propulsion charge was used, there was a valid connection between the statistical clue and the engineering explanation.

Software is available for performing the data analysis tasks described in this chapter (see book homepage: Preface page xvii).

12.9   SOURCES OF DATA

During development, test and failure data will be generated by all tests. As discussed above, the necessary 100 percent FRACAS discipline should ensure that all relevant information is collected and used to drive improvement of the product and process designs.

During manufacture, a 100 percent discipline should also be in place.

Data from in-service use and test is not always as comprehensive, particularly when the use of the product is outside the control of the supplier. Data quality can be improved by:

- Adequate training and management of field service staff and dealers

- Use of warranty repair data, during the warranty period.

Information on the comparative performance of competitors' products can also be obtained, in some cases, via dealers or other support organisations.

Beta testing is a very effective way of obtaining prompt and detailed information in the early stages of a product's service. Beta testing is an extension of development testing (Chapter 6).

REFERENCES

1. Ishikawa, K., 1990, *Introduction to Quality Control*, Chapman & Hall.
2. Ryan, T. P., 1989, *Statistical Methods for Quality Improvement*, Wiley-Interscience.
3. Oakland, J. S., 1996, *Statistical Process Control* (3rd edn), Butterworth-Heinemann.
4. Montgomery, D. C., 1995, *Introduction to Statistical Quality Control*, John Wiley & Sons, Ltd.
5. Deming, W. E., 1986, *Out of the Crisis*, MIT Press.

6. O'Connor, P. D. T., 1995, *Practical Reliability Engineering* (3rd edn), John Wiley & Sons, Ltd.
7. Nelson, W. A., 1982, *Applied Life Data Analysis*, John Wiley & Sons, Ltd.
8. Nelson, W. A., 1990, *Accelerated Testing: Statistical Models, Test Plans and Data Analysis*, John Wiley & Sons, Ltd.
9. Cox, D. R. and Lewis, P. A. W., 1966, *The Statistical Analysis of Series of Events*, Chapman & Hall.

13

Laws, Regulations and Standards

13.1 INTRODUCTION

For most products that are developed and manufactured there are mandatory requirements that must be complied with before they can be sold. These are laid down in laws and regulations and in international, national and market-related standards. Market-related standards are created by agencies and customers such as military, aviation, space, etc. It is obviously necessary that the test programme takes account of all laws, regulations and standards that apply to the product and its markets, and that the necessary testing is planned and integrated into the development and manufacturing test programme.

This type of mandatory testing is generally, but not always, in the category of tests in which we do not want failures to occur (Chapter 6). If we submit a new product to a mandatory test in order to qualify for market acceptance, we hope and expect that it will pass. Therefore, we would not subject it to conditions in excess of the mandatory requirement. However, the margin between the design capability and the required performance must be sufficient to ensure confidence in passing, and in many cases the only way to provide this is to perform earlier tests using the principles of stress acceleration described in Chapter 6.

13.2 LAWS AND REGULATIONS

The laws and regulations that have been created in relation to product engineering relate mostly to safety. There are also laws and regulations covering other aspects, such as environmental impact, maintenance, electromagnetic compatibility, etc., some of which have general applicability and others are specific to particular applications such as public transport. Countries and trading blocs create laws and regulations, so it is necessary to take account of these when designing, developing and manufacturing products for sale and use.

13.2.1 Safety and product liability

The laws and regulations with respect to safety are usually the most important and often the most difficult to deal with. In most countries laws of *product liability* (PL) govern the responsibilities of suppliers in relation to accidents, injuries and deaths caused by the product. Prior to the early 1960s a person who was injured by a product had to prove that the supplier was negligent. This was the principle of *caveat emptor* ('let the buyer beware'). Legislation in the USA replaced this with the principle of *caveat venditor* ('let the seller beware'), or *strict liability*, so that the manufacturer must prove that he has taken all possible steps to prevent the injury. This includes misuse of the product, and the liability extends over the whole product life. In the USA the damages that can be claimed are unlimited, and can include punitive awards.

The law in Europe is similar. The Product Liability Directive (1985) introduced the principle of strict liability, and this has been enacted in the European Community states, for example in the UK by the Consumer Protection Act (1987). The only significant differences from the situation in the USA are that awards are commensurate with the actual damage suffered by the plaintiff, and there are no punitive awards. Nevertheless, the costs of settlements due to injuries or deaths can be very high.

PL legislation has also been introduced in Japan and in other countries.

There are also laws and regulations covering specific safety aspects, such as protection against electric shock, vehicle crashworthiness, testing of pressure vessels, flammability, and radiated emissions. These vary from country to country, and are subject to change.

Therefore product planning must include consideration of all of the general (perhaps even unforeseen) risks as well as the specific risks. Planning for the latter is relatively straightforward: we can know what the requirements are, and we can analyse the design and test the product to ensure compliance. However, dealing with general safety aspects is much more difficult and uncertain. We must consider all of the ways by which the product can present hazards and how they can be eliminated or minimised. This includes operating and maintenance, ways that the product might be misused, protection devices, user instructions, warning labels, etc. The only defence against a failure to anticipate the cause of an injury is to be able to prove that the state of scientific and technical knowledge at the time that the product was introduced was not such as to enable the cause to be foreseen.

Safety laws and regulations therefore place a very heavy burden on the design analysis and test programme. We must make every effort to anticipate hazards, using all the knowledge that is available within the state of the art (not just the knowledge that might be held by the project team or consultants) and of how the product might be used and maintained, over its whole life. The design analysis methods described in Chapter 5 should be used as appropriate. The test principles described in Chapter 6 also apply to testing for safety, in particular the use of highly accelerated stresses (HALT/HASS) to highlight potential weaknesses or defects that are not discovered by analysis or by conventional testing.

All safety-related design and test data must be recorded and retained. Such information is normally required to be presented in support of product approvals, and it might also be needed as evidence that appropriate care was taken to avoid hazards if there is litigation over an accident. This can present dilemmas. Suppose, for example, that analysis or test shows a product defect that might cause a hazard, but the likelihood of occurrence is considered to be very low and the cost of mitigation very high. A decision is made to accept the risk. Should we 'lose' the data, or retain it and hope that the hazard does not occur? In fact, the former option would be illegal, and would generate much higher damages if it were to be discovered. The only legal and ethical way of dealing with this kind of situation is to assess the evidence fully and produce a rational and detailed justification for the decision. If at some future time it is necessary to defend against a claim for damages, such evidence might not provide complete immunity but it would be likely to mitigate the award.

If a potential hazard is discovered after a product has been put into service, whether by test or inspection or as a result of an incident in service, action must be taken to warn and protect users. The most common way of doing this is by a *product recall*. Buyers are traced through dealers or retailers and by advertising, so that the affected products can be taken out of use and replaced or modified. This is an expensive process, but essential and potentially much less expensive than doing nothing.

Liability in relation to safety is mostly a corporate responsibility. However, in some cases where the evidence justifies it, individual designers and managers can be sued.

Reference 1 provides a useful source of information on management and law of product safety in design, development and use.

13.3 MAIN REGULATORY AGENCIES

13.3.1 USA

A number of US Federal agencies generate regulations and monitor safety aspects of products in the USA. These include:

- The Consumer Products Safety Commission (CPSC), which sets standards for and regulates general consumer product safety.

- The Federal Aviation Agency (FAA), which regulates commercial aviation and determines safety and other requirements.

- The National Transportation Safety Board (NTSB), which regulates transport system (road, rail, air) safety and investigates accidents.

- The Underwriters' Laboratory (UL), which determines test and measurement

Figure 13.1   UL mark (Courtesy UL)

Figure 13.2   CE mark (Courtesy UL)

methods and provides facilities and certification. The UL mark (Figure 13.1) on a product indicates that it has been demonstrated to comply with the American standards appropriate to it.

Appendix 2 provides a more detailed list.

13.3.2   Europe

The 'CE mark' (Figure 13.2) signifies that a product conforms with all relevant European Commission (EC) requirements, in particular EC Directives. Relevant products, which include most that involve engineering development and manufacture, may not be sold in EC countries unless they carry the CE mark. It is the responsibility of the vendor to determine the relevant requirements, perform the appropriate analyses and tests (or have them performed) and produce a '*Declaration of Conformity*'. The Declaration of Conformity must state:

- Manufacturer's name and address
- European representative's name and address
- Product name, description and model
- List of directives and standards to which it conforms.

When this has been accepted by an appropriate 'notified body' the product is entitled to carry the CE mark. A 'notified body' is a laboratory, certification organisation, etc., that has been accepted by the appropriate national regulatory organisation, and listed in the *Official Journal of the European Communities.*

The rules for conformity assessment permit several alternative routes, or 'modules'. Table 13.1 outlines these. They range from basic production control to full assessment and registration of the 'quality management system' for design, development and production. The selection of the appropriate module(s) and the full details of compliance are complex, so it is important that they are fully understood and applied. They are detailed in the EC Council Decision 93/465/EEC, available from the EC. A copy is included in Reference 1.

The 'quality management system' (QMS) assessment refers to the ISO9000 approach. This is described later.

The European Commission has issued Directives specific to particular technologies and risks. They include Directives on EMC, electrical safety, etc. (see Appendix 2). These must be complied with as appropriate in order for the CE mark to be applied to the product.

13.4   STANDARDS

The range of standards that relate to engineering is very wide, from definitions of fundamental physical units to specifications of materials and components. Standardisation is necessary for technology advancement, for market regulation and for safety. There is a necessity for certain standards, for example to define basic units, to allocate radio frequencies, to set electromagnetic emissions limits, to define colour codes for electronic components and pressurised gas cylinders, etc. Standards on topics such as these are unambiguous, and it is nearly always possible to determine whether a product complies or not, by inspection, measurement or test.

Originally most standards were created nationally, but many are now international. Many national and industry standards are 'harmonised' with international standards to avoid unnecessary differences between them. Quite often this is done by the international standard being issued unchanged to replace the equivalent national or industry standard, but with retention of the national standard number and title, and sometimes with additional specific requirements.

When compliance with any standards is appropriate for any reason, they cannot be ignored. If compliance is not an option, then it must be included in the design specifications and in the test programme.

Customers for engineering products and systems who specify their requirements, as opposed to buying what is on offer from suppliers, usually include full lists of all standards and other requirements that must be complied with. Military equipment contracts normally include such lists. It is good practice to indicate all requirements for compliance and testing, rather than assuming that the supplier

Table 13.1   Conformity Assessment for CE Mark

Module	Design		Production	
	Manufacturer's responsibilities	Notified body actions	Manufacturer's responsibilities	Notified body actions
A (Internal control of production) B (Type examination)	Retains technical documentation. Submits technical documentation. Describes product type.	Intervenes as required. Assesses conformity. Performs tests if necessary. Issues EC-type examination certificate.	Declares conformity. Affixes CE mark. —	Tests specific aspects. Random product checks. —
C (Conformity to type)	As B	As B	Declares conformity with approved type. Affixes CE mark.	Tests specific aspects. Random product checks.
D (Production quality assurance)	As B	As B	Operates an approved quality management system (QMS) for production and testing. Declares conformity with approved type. Affixes CE mark.	Approves QMS. Carries out surveillance of QMS.
E (Product quality assurance)	As B	As B	Operates an approved quality management system (QMS) for inspection and testing. Declares conformity with approved type or to essential requirements. Affixes CE mark.	Approves QMS. Carries out surveillance of QMS.
F (Product verification)	As B	As B	Declares conformity with approved type or to essential requirements. Affixes CE mark.	Verifies conformity. Issues certificate of conformity.
G (Unit verification)	Submits technical documentation.	—	Submits product. Declares conformity. Affixes CE mark.	Verifies conformity with essential requirements. Issues certificate of conformity.
H (Full quality assurance)	Operates an approved QMS for design.	Carries out surveillance of QMS. Verifies conformity of the design. Issues EC design examination certificate.	Operates an approved QMS for production and testing. Declares conformity. Affixes CE mark.	Carries out surveillance of QMS.

will be aware and will comply. Such lists are also helpful to the supplier in test planning.

The most important international and other standards that relate to testing are listed and briefly described in Appendix 2.

13.4.1   International standards

The International Standards Organisation (ISO), based in Geneva, is the main body responsible for generation of worldwide standards in science and technology. The International Electrotechnical Commission (IEC) is the ISO body responsible for standards in electrical and electronics engineering.

13.4.2   NATO standards

Allied Reliability and Maintainability Procedures (ARMP) describe requirements for NATO military projects. Other NATO standards ('STANAGS') cover a wide range of equipment and materials.

13.4.3   USA

The main standards-producing and monitoring agencies in the USA are:

- The American Bureau of Standards (ABS), which is the central agency for setting and co-ordinating standards, including representation on international bodies such as ISO.

- The Department of Defense (DoD), which issues Military Standards, Specifications and Handbooks (MIL-STDs, MIL-SPECs, MIL-HDBKs), several of which have been mentioned in earlier chapters, and are listed in Appendix 2. Many of the standards, specifications and guidelines were cancelled or downgraded in 1995, when the DoD stated its intention to apply 'best industry practice' wherever practicable.

- Others as listed in Appendix 2.

13.4.4   Europe

The European standards body is the European Committee for Standardisation (CEN). The electrical/electronic subsidiary body is CENELEC. European standards are referred to as 'Euro norms' (EN). Reference 2 provides an introduction to European standards and certification, and further details are given in Appendix 2.

13.4.5   UK

The British Standards Institution (BSI) is one of the oldest standards bodies, and originated many standards still used in engineering worldwide. The BSI 'kite-

Figure 13.3   BSI 'kitemark'

mark' (Figure 13.3) indicates conformity with the relevant Standard(s). British Standards that relate to testing are listed in Appendix 2.

The UK Ministry of Defence issues standards ('DefStans') related to military equipment. The most important from the test point of view are listed in Appendix 2.

13.4.6   Other countries

See Appendix 2.

13.5   'GENERIC' STANDARDS

International standards have recently been generated to cover 'generic' topics such as quality, reliability, safety, and environmental management. However, when standardisation efforts go beyond what is really necessary, and into aspects where compliance cannot be determined objectively, problems rather than solutions tend to be generated. A fundamental problem that arises with most of the standards that have been written on generic topics is that they seldom reflect the best modern practice. Inevitably the process of preparing the standards is slow and bureaucratic, and subject to consensus among the committee members. The committee members might not be familiar with, or might represent interests that conflict with, relevant developments and technologies. When the standards are finally published they are very seldom updated to reflect changed conditions or improved methods, so they 'lock in' outdated and inappropriate methods, and discourage experiment and improvement.

The most important generic standards from the point of view of testing are described below.

13.5.1   ISO9000 (Quality Systems)

The international standard for quality systems, ISO9000, has been developed to provide a framework for assessing the *'quality management system'* which an organisation operates in relation to the goods or services provided. The concept was developed from the US Military Standard for quality management, MIL-Q-9858, which was introduced in the 1950s as a means of assuring the quality of products built for the US military. However, many organisations and companies rely on ISO9000 registration to provide assurance of the quality of products and services they buy and to indicate quality of their products and services.

Registration is to the relevant standard within the ISO9000 'family'. ISO9001 is the standard applicable to organisations that design, develop and manufacture products, and recent changes (ISO9000:2000) have eliminated some of the others. We will refer to 'ISO9000 registration' as a general indication.

ISO9000 does not specifically address the quality of products and services, nor does it prescribe methods for achieving quality, such as design analysis, test and quality control. It describes, in very general and rather vague terms, the 'system' that should be in place to assure quality. In principle, there is nothing in the standard to prevent an organisation from producing poor quality goods or services, so long as written procedures exist and are followed. An organisation with an effective quality system would normally be more likely to take corrective action and to improve processes and service, than would one which is disorganised. However, the fact of registration cannot be taken as assurance of quality. It is often stated that registered organisations can, and sometimes do, produce 'well-documented rubbish'. An alarming number of purchasing and quality managers, in industry and in the public sector, seem to be unaware of this fundamental limitation of the standard.

In the ISO9000 approach, suppliers' quality management systems (organisation, procedures, etc.) are audited by independent 'third party' assessors, who assess compliance with the standard, and issue certificates of registration. Certain organisations are 'accredited' as 'certification bodies' by the appropriate national accreditation services. The justification given for third-party assessment is that it removes the need for every customer to perform his own assessment of all of his suppliers. However, a matter as important as quality cannot safely be left to be assessed spasmodically by third parties, who are unlikely to have the appropriate specialist knowledge, and who cannot be members of the joint supplier–purchaser team. The *total quality management* (TQM) philosophy demands close partnership between suppliers and purchasers.

Since its inception, ISO9000 has generated considerable controversy. The effort and expense that must be expended to obtain and maintain registration tend to engender the attitude that optimal standards of quality have been achieved. The publicity that typically goes with initial certification of a business supports this belief. The objectives of the organisation, and particularly of the staff directly involved in obtaining and maintaining registration, are directed at the maintenance of procedures and at audits to ensure that staff work to them. It becomes more important to work to procedures than to develop better ways of working. Some organisations have generated real improvements as a result of certification, and some consultants and certification bodies do provide good service in quality improvement.

The leading teachers of quality management all argue against the 'systems' approach to quality, and most of the world's leading companies do not rely on it. So why is the approach so widely used? The answer is partly cultural and partly coercion.

The cultural pressure derives from the tendency to believe that people perform better when told what to do, rather than when they are given freedom and the necessary skills and motivation to determine the best ways to perform their work. This belief stems from the concept of *scientific management*, as described in Reference 1 of Chapter 14.

The coercion to apply the standard comes from several directions. In practice, many agencies simply exclude non-registered suppliers, or demand that bidders for contracts must be registered. All contractors and their subcontractors supplying the UK Ministry of Defence must be registered, since the MoD decided to drop its own assessments in favour of the third-party approach, and the US Department of Defense has recently decided to apply ISO9000 in place of MIL-STD-Q9858. Several large companies, as well as public utilities, demand that their suppliers are registered. European Community policy for public purchasing is not explicit on ISO9000, and of course there are no such conditions regarding commercial trade, but this has not prevented registration providers in the USA from advertising that ISO9000 registration is 'a condition for doing business in Europe'. As described above, it is possible to conform to the European CE mark requirements without having ISO9000 registration, though having registration permits a manufacturer greater flexibility.

Defenders of ISO9000 say that the total quality management (TQM) approach is too severe for most organisations, and that ISO9000 can provide a 'foundation' for a subsequent total quality effort. However, the foremost teachers of modern quality management all argue against this view. (It is notable that none of these serve on the national or international committees that prepare and 'update' the standard.) They point out that any organisation can adopt the TQM philosophy, and that it will lead to far greater benefits than will registration to the standard, and at much lower costs. The ISO9000 approach seeks to 'standardise' methods that directly contradict the essential lessons of the modern quality and productivity revolution.

It is notable that ISO9000 is very little used in Japan, and then mainly by companies which perceive that it will provide advantages in Western markets, not because they believe that it will lead to improvements in quality. Companies that embrace TQM set standards for product and service quality, internally and from their suppliers, far in excess of the requirements of ISO9000. These are aimed at the actual quality achievements of the products and services, and at continuous improvement in these levels. Much less emphasis is placed on the 'system'.

The recent changes to ISO9000 ('ISO9000:2000'), apart from introducing requirements for 'improvement' and more detailed requirements for analysis, design, development and measurement, do not deal with these fundamental criticisms.

Many books describe ISO9000 and its application: References 3 and 4 are examples. References 5 and 6 present arguments against the approach.

13.5.2 ISO/IEC60300 ('dependability')

ISO/IEC60300 is the international standard for '*dependability*', which is defined as covering reliability, maintainability and safety. It describes management and methods related to these aspects of product design and development. The methods covered include reliability prediction, design analysis, reliability demonstration tests and mathematical/statistical techniques; most of these are described in separate standards within the ISO/IEC60000 series: see Appendix 2. Manufacturing quality aspects are not included. The methods are mostly inconsistent with modern best practice as described in this book. In particular, the sections on reliability testing define rigid environmental and other conditions to be applied, and for pass/fail criteria based on the statistical methods described and rejected in Chapters 4 and 12.

ISO/IEC60300 has not, so far, been made the subject of audits and registration in the way that ISO9000 has.

13.5.3 ISO/IEC61508 (functional safety of electrical/ electronic/programmable electronic safety-related systems)

ISO/IEC61508 is a new standard to set requirements for design, development, operation and maintenance of 'safety-related' control and protection systems based on electrical, electronics and software technologies. A system is 'safety-related' if any failure to function correctly can present a hazard to people. Thus, systems such as railway signalling, vehicle braking, aircraft controls, fire detection, machine safety interlocks, process plant emergency controls and car airbag initiation systems would be included. The standard lays down criteria for the extent to which such systems must be analysed and tested, including the use of independent assessors, depending on the criticality of the system. It also describes a number of methods for analysing hardware and software designs.

The extent to which the methods are to be applied is determined by the required or desired '*safety integrity level*' (SIL) of the safety function, which is stated within a range of 1 to 4. SIL 4 is the highest, relating to a 'target failure measure' of between 10^{-5} and 10^{-4} per demand, or 10^{-9}–10^{-8} per hour. For SIL 1 the figures are 10^{-2}–10^{-1} and 10^{-6}–10^{-5}. For the quantification of failure probabilities, methods such as MIL-HDBK-217 (see Chapters 3 and 4) are recommended. The methods listed include 'use of well-tried components' (recommended for all SILs), 'simulation' (recommended for SIL 2, 3 and 4), and 'modularisation' ('highly recommended' for all SILs). Most of the analysis and test approaches for hardware and software recommended in this book are not included in the standard.

The standard is without any practical value or merit. The methods described are inconsistent with accepted industry practices, and many of them are known only to specialist academics, presumably including the members of the drafting committee. The issuing of the standard is leading to a growth of bureaucracy, auditors and consultants, and increased costs. It is unlikely to generate any improvements in safety, for the same reasons that ISO9000 does not improve quality.

13.6 INDUSTRY/TECHNOLOGY STANDARDS

13.6.1 Aviation and aerospace

The US Federal Aviation Agency (FAA) issues requirements for commercial and private aviation, related to aspects such as safety and *'airworthiness'*. FAA standards are used by most other countries. FAA requirements that must be demonstrated by test include those for jet engine capability for bird ingestion and for containment of failed rotating components such as turbine discs, reliability and safety of flight controls, aircraft structural integrity, etc. In the UK the Civil Aviation Authority (CAA) performs the same functions, and the European Joint Airworthiness Authority (JAA) has recently been formed to harmonise European requirements.

Air Radio Inc. (ARINC) standards apply to electronic equipment in aircraft (*'avionics'*), and govern aspects such as the dimensions of equipment *'black boxes'* (or *'line replaceable units* (LRUs)), connectors, etc.

National Aeronautical and Space Agency (NASA) standards exist for materials, components and systems used in NASA programmes. The European Space Agency (ESA) sets standards for European space programmes.

The major US aviation companies have recently issued AS9000, a version of ISO9000 adapted for quality systems management of aviation equipment suppliers.

13.6.2 Automotive

The Society of Automotive Engineers in the USA sets standards for automotive applications.

The US 'big three' car companies, GM, Ford and Chrysler (before its buyout by Daimler), have issued QS9000, a version of ISO9000 adapted for automotive suppliers. It is notable that this initiative has not been followed by their Japanese competitors, nor by most other international automotive manufacturers.

However, a draft ISO standard on quality systems for automotive suppliers (ISO/DTR16949) has been prepared and is being discussed.

13.6.3 Other industries

Specialised standards have been created for many other industry sectors, such as telecommunications, railway, petrochemical, medical, construction, etc.

13.7 CONCLUSIONS

The laws and standards that will apply to a product, in relation to industrial, national and international markets, form an essential input to any engineering design. The design team must be made aware of these requirements, and tests must be planned to ensure that the product will comply. '*Homologation*' is a term used to define the process of ensuring compliance with relevant requirements across multiple market jurisdictions.

The design analysis and test results must be fully documented and retained. For many products and markets this is mandatory, but it is sound practice in any engineering situation in order to provide evidence of correct practice in the event of recalls, claims or litigation.

It is arguable that laws and regulations such as the EC CE Directive are inconsistent with the 'strict liability' inherent in product liability law. Compliance with bureaucratic, expensive regulations of a generic 'systems' nature can divert management attention from more effective measures, and inadequate practices can be masked. It is important that the regulatory aspects are managed in such a way as not to interfere with or divert from the practical efforts that really influence quality, reliability and safety.

REFERENCES

1. Abbott, H. and Tyler, M., 1997, *Safer by Design, a Guide to the Management and Law of Designing for Product Safety*, (2nd edn), Gower.
2. Rothery, B., 1996, *Standards and Certification in Europe*, Gower.
3. Hoyle, D., 1998, *ISO9000 Quality Systems Handbook*, Butterworth- Heinemann.
4. Sayle, A. J., 1994, *Meeting ISO9000 in a TQM World* (2nd edn), AJSL Publishing.
5. Hutchins, D., 1992, *Achieve Total Quality*, Director Books.
6. Seddon, J., 1997, *In Pursuit of Quality: the Case against ISO9000*, Oak Tree Press.

14

Managing Test

14.1 INTRODUCTION

The test programme for an engineering project must be managed so that testing can be effectively planned and integrated into the phases of concept creation, design, development, manufacturing and support. Testing, particularly during the development phase, is nearly always expensive, difficult and uncertain in terms of the time and other resources needed. However, the potential for adding value to the project by effective testing is also greatest at this stage. The benefits of effective development testing (and the costs of ineffective testing) accrue in the future, maybe years ahead, so a long-term view is essential.

Test programme integration also means that all of the business functions, such as marketing, finance, engineering (design, development test, manufacturing, maintenance), purchasing and support, must be involved. In turn this means that managing testing is an issue for top management: the different aspects of the test programme must not simply be delegated to specialist functions.

All of the suppliers of engineering products and services to the project must also be involved in the test programme, and their efforts integrated. The contribution to in-service costs of engineering systems due to problems and failures caused by suppliers of components and sub-systems is commonly over 50 percent, and often up to 80 percent, so inadequacies in suppliers' test programmes can prove very expensive later.

Test facilities, such as environmental chambers, ATE, etc., are expensive, as is the knowledge and support that are necessary for effective application. Decisions on what facilities are needed require time for execution, and also entail large and long-term uncertainty and risk. Therefore test facilities planning, including decisions on whether to buy or lease, or to utilise external test services, is an essential aspect.

The test requirements for manufacturing and in-service support must also be planned and managed, and information gained during design and development testing must be used to optimise these.

Technology and risk aspects must also be considered. Novelty always presents risks in engineering, so all product requirements and features that are not supported by experience in the planned application must be given priority attention in the test plan. Other risks, such as market reaction to potential failures, business reputation, hazards, etc., must all be taken into account and costed as part of the cost/benefit analysis for the project.

All regulations and standards for testing that apply to the product and its markets must be identified and complied with as appropriate.

In the sections that follow we will look at all of these management aspects in more detail. We will also discuss the wider management aspects of testing as a key element of engineering management and practice.

14.2   ORGANISATION AND RESPONSIBILITIES

There are two basic engineering organisational roles involved in the management of a test programme: the project manager and the manager responsible for test facilities and support. We will refer to the latter person as the Test Manager, and the organisation as the Test Department.

14.2.1   Test department

The Test Department is responsible for ensuring that the appropriate facilities are available and are used effectively to support projects. It must work closely with project teams to determine and plan what facilities are required, and must also lead the long-term strategy for test provision.

Its responsibilities should also include:

- Knowledge of test methods and requirements (customers, regulations, standards, etc.)

- Managing the use of external test facilities when appropriate (see below)

- Test equipment maintenance and calibration (Chapter 11)

- Test training (see below).

Project managers must retain the responsibility for planning and managing testing on their projects. They must work closely with the Test Manager, and with other functions as appropriate, such as manufacturing, quality and support. Few projects have total control over all test resources, since expensive facilities are usually provided to support multiple projects, and project teams seldom contain all the specialist knowledge required for operating them. A typical project team might contain design and development engineers, and they would make use of facilities and staff of the Test Department. There must be close co-operation between the two teams, starting from initial planning and extending throughout the test programme. The boundaries between the project and Test Department teams must be determined on the basis of factors such as:

- What other projects share the facilities (if there is only one project the roles could be merged)

- The testing that is necessary

- The cost of the facilities (initial and operating)

- The need for test facilities that are unique to the project: such facilities could be provided either by the project or by the Test Department.

14.2.2 Design

The design function must integrate the design of the product and of the downstream processes (manufacture, manufacturing test, maintenance). The ways by which the correctness of the product and process designs will be assured must be determined and managed. Engineers experienced in the downstream processes should be involved as members of the design team, and design engineers should be given training and experience on these aspects. This is an aspect of *integrated engineering*.

The methods of design analysis relevant to testing have been described in Chapter 5 and (for electronics design for test) in Chapter 8. All of these activities must be managed as integral parts of the design process, including formal and informal *design reviews*. Formal design reviews should be scheduled to take place at key stages, particularly:

- Specification: to review and agree the requirements, ensure that they are correct, complete and understood, identify risks, priorities, etc., and determine the design analysis methods to be applied. QFD, as described in Chapter 5, can be a valuable input at this stage.

- Pre-test: to review the design and results of design analyses, agree design changes and finalise the development test plan.

- Pre-production: to review the results of test, agree design changes and retest, and finalise production test methods.

14.3 PROCEDURES FOR TEST

It makes sense to have written procedures to provide a framework for planning project test programmes. Good procedures help to prevent wheels being reinvented in terms of good practice, ensure that responsibilities are clearly defined and enable the best practice to be applied on all projects. They also help to ensure that an integrated approach is followed through all project phases. For compliance with quality system standards (ISO9000, etc.; see Chapter 13) it is necessary to maintain written procedures, and to work to them.

However, it is also important to ensure that written procedures do not impose unnecessary constraints or inflexibility. Managers and engineers should be free to apply methods best suited to the problems and priorities of their projects. As far as practicable, therefore, procedures should be written in the form of guidance, and managers and engineers should be suitably experienced and trained to determine their appropriateness. Some procedures must be mandatory: responsibilities must be clearly defined, and there can be no justification for not reporting and analysing failures (FRACAS) or for using uncalibrated test and measurement equipment.

Procedures for test should cover:

- Organisational and project responsibilities

- Analysis and test methods

- Test planning and execution (development, manufacturing, support)

- Failure reporting, analysis and corrective action system (FRACAS) (Chapter 12)

- Project reviews and design reviews

- Integration with production test planning and transition from development to production

- Maintenance and calibration of test equipment

- In-service maintenance (if appropriate).

14.4 THE DEVELOPMENT TEST PROGRAMME

Engineering projects often run smoothly and happily during the concept and design phases, then hit unplanned costs and delays when testing shows up problems. For many projects the development test programme involves the greatest costs, uncertainties and risks. As far as practicable, it is important to determine in advance what testing will need to be performed on what items. This is necessary to enable the resources (money, test equipment, test items, people, time) to be provided, and for project cost and time budgets to be determined and agreed, but the programme should allow flexibility to take account of problems that might arise.

14.4.1 What to test?

The test programme must cover all levels of the system. For typical engineering products the test programme will include individual modules (components, sub-assemblies/sub-systems like pumps, power supplies, software, etc.), and the overall system. Some of the components and sub-assemblies or sub-systems will be designed and provided by the original equipment manufacturer (OEM), others by suppliers. By the end of the test programme we need to be confident that all components and sub-assemblies/sub-systems, as well as the overall system, achieve the requirements. We must obtain this assurance as quickly and as economically as is consistent with the risks of failure and future costs.

Whenever practicable, the first module to be tested should be the output module, followed by those that provide its inputs, and so on backwards through the system. This approach helps to identify possible problems early and can reduce the impact of changes found to be necessary.

14.4.2   How many to test?

From the point of view of obtaining information, particularly in relation to the effects of variations, the more items we can test the better. On the other hand, provision of test items adds to costs, as does testing effort and time, so a balance must be struck between information and cost. It is usually easier and cheaper to test numbers of components than to test systems, and this is one of the reasons why component (and sub-assembly/system) tests should be part of the test programme. There can be no simple answer to the question of optimum numbers of items to test, except that we must think about it, and determine what seems best for the project.

For low-cost items which cost little to test the answer is simple: test as many as can reasonably be done. However, if the component presents low risks, or if adequate information already exists, then it may not be necessary to perform any tests.

Testing must always be performed at the system level. The numbers to be tested will be determined to some extent by how much and what kind of testing is necessary. For example, the test programme for an electronic system might need to include hardware design proving, software development, environmental tests, mandatory testing for electromagnetic compatibility and beta testing by potential customers. To perform all of these tests sequentially on one unit would not normally be sensible, unless only one or a few were to be produced. Depending on cost, project timescales, risks, production plans and other relevant factors, typical quantities of complex systems such as vehicles, electronics systems and spacecraft in the test programme range from one to 10 or more.

The quantities to be tested will also be influenced by the design analyses, particularly FMEA/FTA.

14.4.3   System levels for test

Testing can be performed at different system levels, from individual components to sub-assemblies/sub-systems and to the overall system.

The advantages of testing components and other lower-level items are:

• They can be tested at lower costs than by testing them as part of higher-level testing, because they are smaller and cheaper. Therefore more items can be tested.

- It is easier and cheaper to apply high stresses, since the items are not protected by other parts of the system such as vibration isolation, thermal inertia, electrical protection, etc.

- Testing can be performed much earlier than system-level testing, so that if problems are highlighted they can be solved before they impact on the much more expensive system tests, and with less delay to the project.

- Such testing can be used as part of the selection process for lower-level items, to determine the most suitable for the project.

On the other hand, lower-level testing does not provide assurance that the items will work together in the system, since it does not test the interfaces between them and other parts of the system. Therefore system-level testing is always necessary. However, as far as practicable we should seek to find and fix all lower-level problems before the system tests.

14.4.4   Testing purchased items

Items that are bought in must be considered in the test programme. The extent of testing should be based upon:

- Existing knowledge, provided in supplier data, past experience of use, etc.
- Risks in relation to technology, novelty, costs, effects of failures, etc.
- The testing performed by the supplier, and knowledge of the results.

For every bought-in item a considered judgement should be made. For example, if a standard electronic component is being applied within the stress and other ratings specified in the manufacturer's data, and its application is not critical, then further testing would not be justified. On the other hand, if, as is the case with some applications, the temperature could exceed the rated values, and the application is critical in terms of cost of failure, then it would be prudent to run high-temperature accelerated tests to validate the selection. Whenever possible the component manufacturer should be advised of the planned tests and invited to advise and assist. In most cases the component manufacturer will be better qualified and equipped to perform any extra testing considered necessary, but this is by no means always the case. However, some manufacturers are not prepared to do so.

Once a particular manufacturer's component is selected for an application on the basis of tests performed by or specified by the user, only that manufacturer's component should be used. Also, it is prudent to repeat the tests from time to time or to obtain the manufacturer's assurance that no changes have occurred in the processes that might invalidate the original test results.

14.4.5 Hardware allocations to test

The overall test programme will consist of a number of different tests, all with particular objectives. These objectives typically include basic performance, reliability/durability, safety, and software development. The hardware items, at all levels, that will be made available for test must be allocated to the different tests. This is particularly important at the higher sub-system and overall system levels, when test items are more expensive and fewer are available. For example, in a large project such as a military system or an aircraft, only four to eight prototypes might be available, and there are similar restrictions on many projects. One way of maximising the possibilities of discovering problems early within such constraints is to move prototypes between tests, so that, for example, more than one is used for each of the test regimes (performance, software, reliability, etc.). This approach helps to maximise the chances of detecting problems that are due to variations (tolerances, etc.).

14.4.6 Test methods

The methods available for development testing have been described in earlier chapters. The test programme must make the best use of all appropriate and available methods, and must not constrain the project team to using only particular defined methods. An old Chinese proverb states *'in my garden grow many beautiful flowers'*. This is a good motto to follow in relation to the methods available for test, as well as for upstream methods of design analysis as described in Chapter 5. Test methods (and other aspects of test, such as quantities, stresses, etc.) must not be based on tradition or on horse-trading between vested interests, but on knowledge of the methods in relation to the project technologies, opportunities and risks.

The technology, costs, risks and mandatory requirements will determine most of the methods that must be applied, such as for vehicle handling, aerodynamic performance, electronic compatibility, etc. To some extent the facilities that are available will also influence the methods to use. However, the two aspects of greatest uncertainty are usually the capability to withstand stresses and the effects of variation. Of course these combine, and can have both short and long term effects, as described earlier. Therefore we must consider the use of the methods appropriate to stresses and variation. These are primarily HALT and the Taguchi approach, as described in Chapter 6.

14.4.7 Development test economics

The starting point for planning development testing should be the costs of failure to achieve the project requirements of performance, reliability/durability, safety and regulatory compliance. The cost of testing to provide assurance of achievement can then be evaluated as an investment to add value, not simply as a burden. The concept of *value added testing* is particularly powerful when applied to development, since the cost of design errors or shortfalls is usually much higher than the costs of manufacturing problems. The ×*10 rule* is often quoted: a problem will cost a factor of 10 higher for each project phase at which it is discovered. Thus if a design error is detected as a result of design analysis or review the cost of rectification might typically be $1000. If the same problem is discovered during the test phase the cost might be $10 000, during manufacture $100 000 and after introduction to service $1m. Situations in which these ratios have been much higher are common.

The author has never found a company or a project where it was considered in retrospect by those involved that too much was spent on a development test programme. Hindsight almost always indicates that more and better testing would have been a worthwhile investment, because quite often the consequences of inadequate testing are very expensive indeed.

There are some development test methods that are expensive yet relatively ineffective. In general, any tests in which equipment is operated for long periods of time under 'typical' conditions are wasteful. These include reliability demonstration tests, as described in Chapter 12. Quite often such tests are part of the tradition of an industry or a company. An interesting counter-example is that of the railway rolling stock industry, for which, until recent denationalisation and deregulation in Europe, there has been an historical tradition of almost no development testing before new trains were put into service.

14.4.8   Use of external test facilities

A wide range of test facilities are available to support product development. Some are run as test businesses, others as services offered by companies or other organisations that have spare time on facilities that they own. The available facilities cover the range of test technologies and industry applications. Generally speaking, they provide a service in relation to development testing and for regulatory compliance testing, but not for production.

Test providers have the necessary specialist staff and expertise to conduct the appropriate tests. They will normally guarantee confidentiality, and their work will be competent and unbiased. It is often cheaper to use external test providers than to set up and run internal facilities. Decisions in this respect must be made on the basis of expected facility utilisation, costs, and the extent to which the test equipment and methods represent core technologies for the business. (A core technology is one that provides a competitive advantage, and that cannot be purchased effectively from competing suppliers (Reference 1).)

Advantages of in-house testing for development are:

- Core technologies retained

- Opportunities for more involvement by designers in testing

- Possibly better confidentiality

- Greater flexibility

- Usually less expensive, particularly if HALT/HASS methods are applied.

Disadvantages are:

- Capital tied up

- Possibility that the latest test technologies might not be made available, due to budget or other constraints.

For many projects a balanced use of internal and external facilities is appropriate. The test procedures and project test plans (see below) should cover how this aspect will be managed.

Information on test service providers is available on the book homepage (see Preface page xvii).

14.4.9 Regulations and standards

The regulations and standards relevant to testing were described and discussed in Chapter 13. It is essential that all of the relevant requirements are known, understood and included in test programmes, procedures and training. This aspect is appropriate to the responsibilities of the Test Manager.

It is important that any mandatory tests that are considered inappropriate or not cost-effective are dealt with correctly. There must be evidence that the tests are performed, in the form of procedures, responsibilities and test records. However, such tests should be minimised in terms of cost and trouble, so a good understanding of exactly what is required is important.

14.5 THE PROJECT TEST PLAN

A *test plan* should be prepared for every project. This should be the responsibility of the project manager. For some kinds of project the test plan is a contractual requirement; for example, the US Department of Defense requires a *test and evaluation master plan* (TEMP) to be submitted and complied with.

The test plan should include:

- The performance and other requirements that are to be achieved. This must include any mandatory requirements such as compliance with contracts, standards or regulations, as well as requirements for reliability, durability and safety.

- The important failure modes that should be prevented from occurring

- The design and design analysis inputs to the test programme (Chapter 5)

- The test (and other methods if appropriate) that will be used to ensure that the requirements are achieved (Chapters 6–9)

- Items to be tested (components, sub-systems, system), and hardware allocations to the different tests when appropriate

- The test requirements for suppliers to the project

- Integration of test methods through project phases

- Responsibilities for performance of tasks and for providing support

- Schedules, linked to the main project schedule.

The test plan should be linked to and consistent with any other relevant project plans, such as for reliability, quality, safety and manufacturing. However, there should be only one test plan for development, not separate plans for testing for performance, reliability, safety, etc.

The test plan should be based upon, and should make reference to, any relevant contracts, written procedures, etc. as described earlier. For example, some customers require a test plan, or a test and evaluation plan (Chapter 1), to be prepared and accepted.

Figure 14.1 is a flow chart for the development test programme, showing the aspects that should be considered and the appropriate methods for the analysis and test tasks to be performed. An outline development test plan is presented in Appendix 3.

14.6   MANUFACTURING AND MAINTENANCE

Manufacturing test methods and economics have been described in earlier chapters. The objectives of manufacturing tests are fundamentally different from those of development testing. Manufacturing tests must show which items are not 'good', without damaging good ones. Therefore the question of the quantities to be tested is also answered differently. We must test all or none, or a (non-statistical) sample, as explained in Chapter 12.

The use of accelerated stresses must be considered, particularly HASS.

The principles and methods described in Chapter 11 should be applied to in-service testing.

Figure 14.2 is a flow chart for the manufacturing and in-service test programmes, showing the aspects that should be considered and the appropriate methods for the analysis and test tasks to be performed. An outline production test plan is presented in Appendix 4.

14.7   TRAINING AND EDUCATION FOR TEST

Design engineers and engineering managers are generally familiar with basic deterministic aspects of design and functional test. However, they often lack the

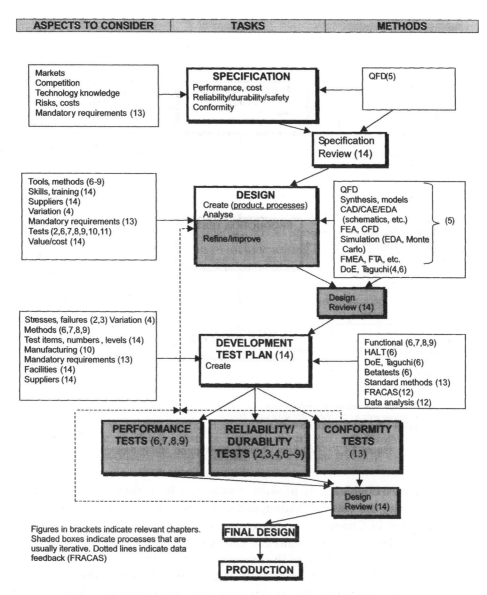

ASPECTS TO CONSIDER	TASKS	METHODS

Markets
Competition
Technology knowledge
Risks, costs
Mandatory requirements (13)

SPECIFICATION
Performance, cost
Reliability/durability/safety
Conformity

QFD(5)

Specification
Review (14)

Tools, methods (6-9)
Skills, training (14)
Suppliers (14)
Variation (4)
Mandatory requirements (13)
Tests (2,6,7,8,9,10,11)
Value/cost (14)

DESIGN
Create (product, processes)
Analyse

Refine/improve

QFD
Synthesis, models
CAD/CAE/EDA
 (schematics, etc.)
FEA, CFD
Simulation (EDA, Monte
 Carlo)
FMEA, FTA, etc.
DoE, Taguchi(4,6)
(5)

Design
Review (14)

Stresses, failures (2,3) Variation (4)
Methods (6,7,8,9)
Test items, numbers , levels (14)
Manufacturing (10)
Mandatory requirements (13)
Facilities (14)
Suppliers (14)

**DEVELOPMENT
TEST PLAN** (14)
Create

Functional (6,7,8,9)
HALT(6)
DoE, Taguchi(6)
Betatests (6)
Standard methods (13)
FRACAS(12)
Data analysis (12)

**PERFORMANCE
TESTS** (6,7,8,9)

**RELIABILITY/
DURABILITY
TESTS** (2,3,4,6–9)

**CONFORMITY
TESTS**
(13)

Design
Review (14)

Figures in brackets indicate relevant chapters.
Shaded boxes indicate processes that are
usually iterative. Dotted lines indicate data
feedback (FRACAS)

FINAL DESIGN

PRODUCTION

Figure 14.1 Test flow (design and development)

knowledge, experience and insights relevant to designing in relation to variation
and reliability/durability (Chapter 4) and accelerated test, particularly HALT
and HASS (Chapters 6 and 10). There is also widespread ignorance of many of
the other aspects of test covered in this book. Therefore there is nearly always a
great potential for improving the capability of design engineers and teams to 'get
it right the first time'. Fewer mistakes and oversights during the design phase can

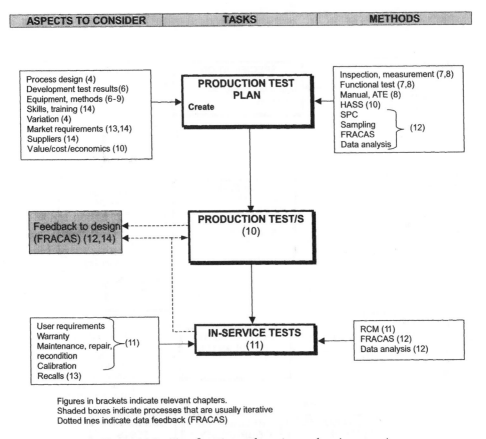

| ASPECTS TO CONSIDER | TASKS | METHODS |

Figures in brackets indicate relevant chapters.
Shaded boxes indicate processes that are usually iterative
Dotted lnes indicate data feedback (FRACAS)

Figure 14.2 Test flow (manufacturing and maintenance)

greatly improve downstream product economics, by orders of magnitude in comparison with the costs of finding and correcting problems later.

As explained in Chapter 6, training designers in test methods and exposing them to the testing of their designs can generate knowledge and experience that leads directly to improvement of their capabilities to create correct designs in the future. This applies to all testing, but HALT provides the extra benefit of generating results very quickly, so that designers can participate effectively and learn quickly. Such training and experience generates continuous improvement (*kaizen*) of design capability.

Testing is of enormous importance to all engineering. Nevertheless, it is a stark fact that it is an aspect that is very little taught, particularly as part of engineering degree courses. When testing is taught it is mostly limited to specialised aspects, such as fatigue testing as part of a mechanical engineering course or software testing on a computer science course. There is little postgraduate teaching, on either higher degree or short courses. The teaching of engineering theory and design is essential, but not to the exclusion of the equally important and often more difficult work of test. It is interesting that nearly all engineering courses include elements of management, on the assumption that all of the engineers they teach will become managers at some stage. However, not all will, and if they do it will not usually happen for some years after graduation. Management teaching, even when not done very well (as is unfortunately too often the case), confers 'status' on the course, but practical aspects like testing are perceived as being insufficiently academic. However, any engineer who does not appreciate how designs and products should be tested is handicapped from the day he or she starts work.

The facts that this is the first book to appear on the subject, and that Reference 1 is the only book on engineering management that covers testing, are also indicative of the scant attention that has traditionally been paid to the subject.

The providers of engineering degree courses should review this aspect of their curricula, and should take action to include testing as appropriate to the courses. For all courses there should be an introduction to the basic philosophy and methods of test, in development, production and use. Specialist technology aspects related to particular courses should be included as appropriate. Universities and other appropriate institutions should also consider setting up short courses in testing, for managers and for practising engineers. The major professional engineering institutions and societies should also address this topic.

Engineers do learn about test, but mostly on the job. There is nothing wrong with on-the-job learning for practical specialist aspects. However, it seldom covers many of the important and wider aspects discussed in this book, such as the links between design analysis, development testing and manufacturing test, accelerated test methods, testing for variation and regulatory aspects. The employers of engineers should consider what internal training would be appropriate in relation to their products and markets.

The training and development of engineers is discussed in more detail in Reference 1.

14.8 THE FUTURE OF TEST

14.8.1 Virtual testing

Virtual testing is already being applied, most notably in electronics design and for systems involving fluid flow, such as aircraft and combustion systems, as described in earlier chapters. With the rapid developments in computer-aided engineering, including virtual reality software, we will see more use of such systems to supplement or replace traditional development testing. There will be further development, including virtual 'failure creation' (fracture, bearing seizure, parameter change, etc.) as a result of the applied 'tests'. Such tests will also be integrated with actual tests, an extension of present 'hardware in the loop' tests. However, virtual testing will be only as effective as the models allow. There will always be a need for some real testing using actual hardware.

14.8.2   Intelligent CAE

CAE software that can simulate multiple technologies and integrate aspects such as performance, spatial properties such as movements and clearances, strength and stress analysis, and variation will greatly increase our ability to create correct initial designs and to reduce the time and cost of test. More advanced and integrated EDA software will continue to be developed to create, analyse and 'test' the very complex electronics designs of the future.

CAE software will be developed to be more 'intelligent' and integrated by further development of seamless transitions across technologies (electronics, mechanics, materials, fluid dynamics, software, etc.). It will include knowledge and features such as design rules, guidelines, component and material selection information, failure modes and effects analysis, etc., and databases of past experience.

A design station of the future might be analogous to a modern word processor. Design errors would be automatically highlighted, and 'help' menus would be available to guide the designer to the correct approaches. A 'thesaurus' would be available to assist in selection of components, materials and sub-systems, with links to Internet sites for more information, order placing, etc. An 'insert' function would allow components and sub-system drawings and details to be added. 'Tools' would include analysis of strength, stress, variations, etc., as well as designing and running virtual tests. Some of these features are already available in EDA software, but have yet to be extended into the more varied world of design of multi-technology products.

Engineering design and development is a 'fuzzy' process, except in some areas like purely digital electronics. CAE systems of the future might include fuzzy logic for tasks such as tolerancing, reliability analysis, etc.

Ultimately the inputs to system design might consist of a detailed performance specification, and the CAE will do the rest! However, this is a distant scenario, and is likely to be achieved only in limited applications. Software will not replace the need for hardware testing in our lifetimes.

14.8.3   The Internet

The Internet is quickly becoming an essential tool and information source for engineering design and management. Together with company intranets, it enables test and other data to be accessed and transmitted more rapidly and cheaply than ever before. Test results, from development, manufacturing or maintenance and calibration, can be collected and uploaded to another site for analysis, and the analysed results downloaded in seconds, or even automatically. Designers, test engineers, suppliers, etc. can share and discuss information, including pictures and videos, with no delay and at very low cost.

Already there are several Internet sites that provide information and support on testing. See book homepage (Preface page xvii).

Internet-based design is a growth area. The features discussed in the earlier paragraphs will be provided to a large extent via the Internet, rather than on stand-alone CAE systems.

14.8.4 Test hardware

Built-in test capabilities will increase, spearheaded by microelectronics and nano-technology, such as for sensors and mechanisms. More systems and kinds of products will contain features for continuous monitoring or remotely activated functional tests. An interesting recent example is the *sentient instrument control-ler*™ (Management Sciences, Inc.) developed for continuous monitoring of electrical cabling in systems such as aircraft, factories, etc., to monitor electrical parameters and to detect and report malfunctions such as breaks, shorts, overheating, etc. The *bluetooth* short-range wireless interface standard will be applied to test instruments and to units to be tested, enabling units to run built-in test and to communicate the results to instruments or data systems without the need for cable connections.

14.8.5 Teaching testing

Testing should be brought into mainstream engineering curricula. The engineering institutions and societies should insist on testing becoming an essential part of the courses they accredit. As a result, practising engineers and engineering managers would be better trained and more experienced in test. Not least, it is hoped that many thousands of copies of this book will be bought.

14.9 CONCLUSIONS

Testing is usually the most expensive and difficult task in the development of engineering products and systems. It is in the test phase that the errors and oversights that are inevitable in nearly all modern engineering design should be detected and corrected, rather than coming home to roost after introduction to service. Development test programmes always entail a large amount of uncertainty, because we cannot predict what problems might occur or how serious they might be. We can greatly reduce the potential for errors and oversights by ensuring that design teams are better trained and more experienced. We can also detect and correct design problems by effective use of design analysis and review methods. However, it will nearly always be necessary to perform tests to ensure that the design of the product and of the processes is acceptable in terms of performance, reliability, durability, safety and, in many cases, regulatory compliance.

Therefore development testing must be managed as an integral aspect of the whole product process. Design and test must be closely integrated from the earliest stages, and designers should be active participants in the analysis and testing of their designs. Suppliers' test programmes and methods must also be managed as part of the overall project.

Testing is also an integral part of the manufacturing process, and often of in-service support and the methods to be applied must therefore be designed and tested during the preceding phases. Design teams should be aware of the relevant manufacturing and maintenance test technologies and methods.

The effective management of testing is inseparable from the wider aspects of managing engineering, and the fundamental principles of good management apply. References 1–5 cover the wider aspects of engineering management (though only Reference 1 actually covers test).

The technology of test is complex and dynamic. Few engineers are adequately trained or experienced, and inappropriate practice is very common. Some examples were described in Chapter 1. This is obviously a problem, but it can also be taken as an opportunity to improve. In many cases improvements in test management and methods, as described in this book, can generate large savings and commercial advantage in terms of engineering and manufacturing produc-tivity and product reputation.

Whilst development and manufacturing testing is expensive, insufficient or inadequate testing can be far more costly later, often by orders of magnitude. Therefore the test programme must be planned and financed as a long-term investment, not merely as a short-term cost. This can be a difficult concept to sell, particularly as so many organisations are driven by short-term financial measures like end-of-year profits, dividends and stock options. Engineering as well as commercial experience and judgement must be applied to the difficult and uncertain business of test. Managers at all levels and in all contributing functions must appreciate the concept that *test is an investment that must be planned*, and that can generate very large returns. *Test adds value.*

REFERENCES

1. O'Connor, P.D.T., 1994, *The Practice of Engineering Management*, John Wiley & Sons, Ltd
2. Drucker, P.F., 1955, *The Practice of Management*, Heinemann.
3. Clausing, D., 1994, *Total Quality Development*, ASME Press.
4. Bergman, B. and Klefsjo, B., 1994, *Quality: from Customer Needs to Customer Satisfaction*, McGraw-Hill.
5. Conti, T., 1993, *Building Total Quality*, Chapman & Hall.

Appendix 1

Acronyms

ABS	American Bureau of Standards
AC	Alternating current
A/D	Analogue-to-digital
AGREE	Advisory Group on Reliability of Electronic Equipment
ANOVA	Analysis of variance
AOI	Automatic optical inspection
AQL	Acceptable quality level
ARINC	Air Radio Inc.
ASIC	Application-specific IC
ASME	American Society of Mechanical Engineers
ATE	Automatic test equipment
ATLAS	Abbreviated test language for all systems
ATPG	Automatic test program generation
AVO	Amps, volts, ohms (meters)
AXI	Automatic X-ray inspection
BGA	Ball grid array (IC Package)
BHN	Brinell hardness number
BIST	Built-in self-test
BIT	Built-in test
BS(I)	British Standards (Institution)
CAA	Civil Aviation Authority (UK)
CAD	Computer-aided design (or drafting)
CAE	Computer-aided engineering
CE (Mark)	European Commission
CEN	European Committee for Standardisation
CENELEC	The electrical/electronic subsidiary body of CEN
CERT	Combined environment reliability testing
CFD	Computational fluid dynamics
CMM	Co-ordinate measurement machine
CMOS	Complementary metal-oxide silicon (IC)
C_p, C_{pk}	(Statistical) process capability
CPSC	Consumer Products Safety Commission (USA)
CRT	Cathode ray tube
CSP	Chip scale packaging (IC package)

D/A	Digital-to-analogue
DAQ	Data acquisition
DC	Direct current
DefStan	Defence standard (UK)
DIP	Dual in-line package (IC package)
DMM	Digital multimeter
DoD	Department of Defense (USA)
DoE	(Statistical) design of experiments
EC	European Commission
EDA	Electronic design automation
EM	Electromigration
EN	Euro norm (European standard)
EO	Electro-optical
EOS	Electrical overstress
ESA	European Space Agency
ESD	Electrostatic damage
ESS	Environmental stress screening
FAA	Federal Aviation Agency (USA)
FEA	Finite element analysis
FIFO	First-in first-out (logic)
FMEA	Failure modes and effects analysis
FMECA	Failure modes, effects and criticality analysis
FMVT	Failure mode verification testing ($^{®}$ Entela Corp.)
FPGA	Field programmable gate array
Fpmh	Failures per million hours
FRACAS	Failure reporting, analysis and corrective action system
FT	Functional test(er)
FTA	Fault tree analysis
GPIB	General purpose interface bus
HALT	Highly accelerated life test
HASS	Highly accelerated stress screen
(H)AST	(Highly) accelerated stress test
HAZOPS	Hazard and operability study
HLL	High level language
HOL	High order language
IC	Integrated circuit
ICT	In-circuit test
IEC	International Electrotechnical Commission
IEST	Institute of Environmental Sciences and Technology
IID	Independently and identically distributed (variable)
ISO	International Standards Organisation
JAA	Joint Airworthiness Authority (Europe)
LED	Light-emitting diode
LIFO	Last-in first-out (logic)
LR	Loading roughness
LRU	Line replaceable unit

LSA	Load–strength analysis
LSI	Large scale integration (IC) (up to 10 000 logic gates)
LSSD	Level sensitive scan design
LTPD	Lot tolerance percent defective
MAST	Multi-axis shake table
MCM	Multi-chip module (IC package)
MDA	Manufacturing defects analyser
MMI	Man-machine interface
MMIC	Microwave monolithic IC
MSI	Medium scale integration (IC) (up to 1000 logic gates)
MTBF	Mean time between failures
MTTF	Mean time to failure
NASA	National Astronautics and Space Administration (USA)
NATO	North Atlantic Treaty Organisation
NDT	Non-destructive test
NFF	No fault found
NMR	Nuclear magnetic resonance
NTSB	National Transportation Safety Board (USA)
NVH	Noise, vibration and harshness
OEM	Original equipment manufacturer
PCI	PC interface (bus)
Pdf	Probability density function
PED	Plastic (epoxy) encapsulated device (IC package)
PFMEA	Process FMEA
PGA	Pin grid array (IC package)
PL	Product liability
PLC	Programmable logic controller
PLD	Programmable logic device
PRST	Probability ratio sequential testing
PSD	Power spectral density
PWM	Pulse-width modulated
PXI	PC extensions for instruments (bus)
QFD	Quality function deployment
QFP	Quad flat-pack (IC package)
QMS	Quality management system
RCM	Reliability centred maintenance
RDT	Reliability demonstration test
RGM	Reliability growth monitoring
RH	Relative humidity
RP	Rapid prototyping
RTOK	Retest OK
SA	Sneak analysis
SAE	Society of Automotive Engineers
SD	Standard deviation
SIL	Safety integrity level
SM	Safety margin

SMT	Surface mount technology (electronics)
SOAP	Spectrometric oil analysis programme
SOIC	Small outline IC
SPC	Statistical process control
SSI	Small scale integration (IC) (up to 100 logic gates)
STANAG	Standardisation agreement (NATO standard)
STRIFE	Stress-induced failure environment
TCE	Thermal coefficient of expansion
TDDB	Time dependent dielectric breakdown
TDR	Time domain reflectometer
T&E	Test and evaluation
TEMP	Test and evaluation master plan
TQM	Total quality management
TSA	Time series analysis
UL	Underwriters' Laboratory (USA)
ULSI	Ultra large scale integration (IC) (more than 100 000 logic gates)
UTS	Ultimate tensile strength
UUT	Unit under test
UV	Ultraviolet (radiation)
VAST	Versatile avionic shop test
VHDL	VLSI hardware description language
VLSI	Very large scale integration (IC) (up to 100 000 logic gates)
VR	Virtual reality
VXI	VLSI extensions for instruments (bus)

Appendix 2
Testing Regulations and Standards

The lists in this appendix are of necessity incomplete. There are many standards-setting organisations and standards that are specific to particular industries, types of product, national markets, etc., and changes are frequent. The lists are a selective guide to those that are probably the most widely known and used. For particular situations it is essential that the full details of applicable requirements are determined.

STANDARDS ORGANISATIONS (GENERAL)

International Organisation for Standardisation (ISO) (http://www.iso.ch).

International Electrotechnical Commission (IEC) (http://www.iec.ch). The electrical/electronic section of ISO.

US National Bureau of Standards (NBS) (http://www.nbs.org). Co-ordinates the production of American standards.

European Committee for Standardisation (CEN). European standards are prefixed EN ('Euro norm').

European Committee for Electrotechnical Standardization (CENELEC).

European Commission (EC). The EC issues directives related to compliance with standards.

British Standards Institution (BSI) (http.//www.bsi.org.uk). The BSI co-ordinates the production of UK standards. British standards are prefixed BS.

UK Accreditation Service (UKAS) (http://www.ukas.com). UKAS is responsible for the accreditation of laboratories offering measurement and calibration services. UKAS also provides accreditation for ISO9000 certification organisations.

German Association for Electronics and Information Technology (VDE) (http://www.vde.de). VDE produces standards for electronics and IT, provides accreditation services, and operates test laboratories. The VDE mark is affixed to equipment which complies with relevant VDE standards.

(All other industrialised countries have national standards organisations, which contribute to international standardisation work. However, only a few maintain standards that differ significantly from international or other major national standards.)

INDUSTRY-BASED STANDARDS ORGANIZATIONS
Mechanical, systems
American Society of Mechanical Engineers (ASME) (http://www.asme.com).

American Society for Testing and Materials (ASTM) (http:www.astm.com).

Electrical, electronic
Institute of Electrical and Electronic Engineers (IEEE) (http://www.ieee.org).

Military
US Department of Defense (Military standards, handbooks and specifications: MIL STDs, MIL HDBKs, MIL SPECs).

North Atlantic Treaty Organisation (NATO Standards: STANAGS).

UK Ministry of Defence (Defence Standards: DefStans).

Aviation, aerospace
International Civil Aviation Organisation (ICAO).

Air Radio Inc. (ARINC). Produces standards for aviation electronic systems (avionics).

National Astronautics and Space Administration (NASA).

Federal Aviation Administration (FAA) (USA).

Civil Aviation Administration (CM) (UK).

European Space Agency (ESA).

Automotive
Society of Automotive Engineers (SAE) (USA).

ISO/IEC STANDARDS

ISO57025 Accuracy (Trueness and Precision) of Measurement Methods and Results. Describes methods and provides guidance on the quality of measurements, not specific to any technologies.

ISO/IEC60068 Environmental Testing. General guidance and descriptions of environmental test methods for a range of conditions. Similar to US MIL-STD-810, and replaces BS2011.

ISO/IEC60512 Electromechanical Components for Electronic Equipment: Basic Testing Procedures and Measuring Methods.

IEC61000 Electromagnetic Compatibility (series). Describes EMC environments, limits, test and measurement methods, and installation guidelines.

ISO9000 (series). Quality Assurance. *Discussed in Chapter 13.*

ISO/IEC60300 Dependability Management. Describes methods for reliability, maintainability and safety analysis and test. Some parts are still being drafted. *Discussed in Chapter 13.*

ISO/IEC60605 Equipment Reliability Testing. *See Chapter 13.*

ISO/IEC80721 Classification of Environmental Conditions. Describes 'standard' environmental conditions for testing a range of environments, such as static protected, mobile, storage, etc. Replaces BS7527. *See Chapter 13.*

ISO/IEC61163 Reliability Stress Screening. *See Chapter 13.*

lSO/IEC61069 Industrial Process Measurement and Control: Evaluation of System Properties for the Purpose of System Assessment. Describes methods for assessing system 'functionality', performance, 'dependability', 'operability' and safety. *See Chapter 13.*

ISO/IEC61508 Functional Safety of Electrical/Electronic Programmable Electronic Safety-Related Systems. *Discussed in Chapter 13.*

ISO/IEC17025 Operation of Test Laboratories. A new standard in course of preparation, to describe procedures and methods for the operation and assessment of test and calibration laboratories. It will replace the European EN45001 series (see below). The management system requirements are based on ISO9000.

US STANDARDS

The ASME pressure vessel codes, which include test methods, are applied worldwide.

The IEEE have issued many standards related to testing electrical and electronics systems. These include:

- IEEE488 Test Databus
- IEEE1149.1 Boundary Scan.

US Military standards

Many of the US Military standards, specifications and handbooks were cancelled or downgraded in 1995, when the DoD stated its intention to apply 'best industry practice' wherever practicable. The most important retained documents from the test point of view are listed below (the letter after a reference number indicates the latest issue).

MIL-STD-810E Environmental Test Methods and Engineering Guidelines. Describes the range of test methods and conditions for military equipment. The tests include temperature, vibration, shock, humidity, and a range of other conditions.

MIL-HDBK-781 Reliability Testing for Engineering Development, Qualification and Production. This was downgraded from a standard to a handbook. it describes the engineering and analytical methods to be used to demonstrate reliability (MTBF). *It was described in Chapter 12.*

MIL-HDBK-210C Climatic Information to Determine Design and Test Requirements for Military Systems and Equipment.

MIL-STD-202F Test Methods for Electronic and Electronic Components.

MIL-STD-883D Test Methods and Procedures for Microelectronics. *Described in Chapter 10.*

MIL-STD-461D Requirements for the Control of Electromagnetic Interference Emissions and Susceptibility.

MIL-STD-462D Measurement of Electromagnetic Interference Characteristics.

MIL-E-6051 D Electromagnetic Compatibility Requirements, Systems.

MIL-HDBK-2164 Environmental Stress Screening Process for Electronic Equipment.

MIL-HDBK-2165A Testability Program for Systems and Equipments.

MIL-HDBK-338 Electronic Reliability Design Handbook, 'for guidance only'.

EUROPEAN DIRECTIVES AND STANDARDS

European Commission Directives

The most important EC Directives from the point of view of testing are:

- General Product Safety Directive

- Machinery Directive (89/392/EEC) which covers safety and other aspects of 'machines', which are broadly defined to include products such as power tools, motors, electrical equipment, etc.

- Electromagnetic Compatibility (EMC) Directive (89/336/EEC)

- Low Voltage Electrical Equipment Directive (73/23/EEC)

- Radio Equipment and Telecommunications Terminal Equipment (R&TTE) Directive (1999/5/EC)

- Medical Devices Directive (93/42/EEC).

European standards ('Euro norms')

EN45001/EN45002; EN45003 General Criteria for the Operation/Assessment of Testing Laboratories; Calibration and Testing Laboratories Accreditation Systems. Replace BS7501/7502. Will be replaced by ISO/IEC17025.

EN45014 General Criteria for Suppliers' Declaration of Conformity. *See Chapter 13.*

British Standards (BS)

British Standards have been to a large extent harmonised with international standards, as described above. Another British Standard relevant to testing is:

BS5760 Reliability of Systems, Equipments and Components. BS5760 is a 'guidelines' standard that describes methods for reliability achievement, including design analysis, test and data analysis.

UK Defence Standards (DefStans)

DefStan 00-40 The Management of Reliability and Maintainability, and DefStan 00-41 Reliability and Maintainability Methods. These standards describe the management and methods, respectively, that should be applied to the achievement of reliability, availability and maintainability of UK defence equipment. They were produced to provide a realistic and practical framework. The methods described are generally consistent with those recommended in this book.

DefStan 00-13 Testability. Provides guidelines and assessment methods for the testability of military electronic systems, with the emphasis on in-service test and diagnosis.

Appendix 3
Development Test Plan Example

DEVELOPMENT TEST PLAN

POSITION CONTROLLER

SYSTEM DESCRIPTION

Electro-pneumatic precision positioning system. Specification No. 9999. Drawing No. DDDD.

REQUIREMENTS TO BE TESTED

Operation

Load: 2–5 Kg. Friction and stiction to be determined on test.
Maximum travel: 500 mm. Safety cutout limits at 10 mm beyond max/min travel.
Accuracy: ±1 mm. Overshoot 0 mm.
Operating frequency range: 0–0.5 Hz with 5 Kg load.

Environment

Temperature: 10–40 Deg C, maximum rate of change 5 Deg C/min.
Humidity: 50–100% RH, with condensation.
Vibration/shock: NIL
Electrical power: 240 V ±10 V
Air: 20 bar, filtered to 5 micron.

Durability/reliability

5 years continuous operation, 50 demands/day (100,000 cycles).
Zero safety failures. Not more than 1 operating or accuracy failure per system per year.

Maintenance

> Weekly: accuracy test.
> 6-monthly: examine, clean.
> 12-monthly: replace actuator seals.

Conformity

> UL and CE requirements for EMI/EMC.
> CE Machinery Directive.

ITEMS TO BE TESTED

> Electronic control unit (ECU) (2)
> Servo actuator (SA) (2)
> Feedback potentiometer (FP) (10 each)
> Software (S/W)
> Overall system (SYS) (2)

DESIGN ANALYSIS

The system design will be subjected to the following design analyses:

ECU and software

> Circuit simulation, covering operation and parameter variations. Include input values and components (power, input demands, feedback, limits, etc.).
> EMI/EMC.
> Testability analysis.

Servo actuator

> Strength and fatigue analysis of actuator mounting.

System

> FMEA, FTA.
> System simulation, covering operation, frequency response over load range, parameter variations.
> CE Machinery Directive conformity requirements.

TESTS
ECU

Functional: 2 units.
HALT: Combined temperature, vibration and humidity/condensation. Electrical power variation. 2 units.
EMI/EMC: 1 unit.
Production test development: 1 unit.

Servo actuator (including mounting)

HALT: Combined temperature, vibration and humidity/condensation. Air supply variation (pressure, filtration). Load variation. 2 units.
Durability (wear, fatigue): 200 K cycles at max (5 Kg) load, max frequency. 2 units.

Feedback potentiometer (tests to be performed on 3 competing components)

HALT: Combined temperature, vibration and humidity/condensation. Electrical power variation. 10 units of each.
Durability: 200 K cycles at 40 Deg C. 10 units of each. Test at end for function to specification.

Software

To be tested in accordance with Procedure PPPP. Tests to cover:
- Operating logic.
- Limits (control and safety cutouts, power and air supply, other safety limits).

System

Functional: 2 systems.
HALT: environments as above. 2 systems.
Conformity (CE, UL). 1 system.
Production and maintenance/calibration test development. 1 system.

PROCEDURES AND RESPONSIBILITIES

All analyses and tests to be performed in accordance with relevant procedures.

REPORTING

Reports are to be produced on all analyses and tests, in accordance with Procedure RRRR. All failures are to be reported and investigated in accordance with Procedure FFFF.

Analysis, test and failure investigation reports are to be submitted to and actioned at Design Reviews, in accordance with Procedure DDDD.

SCHEDULE AND HARDWARE ALLOCATION

(Reports: ♦)

MONTH → / TASK ↓	1	2	3	4	5	6	7	8	9
Design analysis	Initial			♦	Repeat	as	required	♦	
ECU F/T			ECU 1,2		♦				
ECU HALT					ECU 1	ECU 2	♦		
EMI/EMC						ECU 1	♦		
ECU Prod/maint					ECU 2	♦			
SA HALT, Durability		HALT ♦	Durability				♦		
FP HALT, Durability			HALT ♦	Durability				♦	
S/W Test									
SYS F/T				SYS 1	♦	SYS 2	♦		
SYS HALT					SYS 1	♦	SYS 2	♦	
SYS Prod/maint						SYS 1	♦		
Conformity							SYS 1	♦	
Re-tests*							*As	required	
Design Review		(1)		(2)				(3)	
Build			SYS 1		SYS 2				
Start production									

Appendix 4

Production Test Plan Example

PRODUCTION TEST PLAN

POSITION CONTROLLER

SYSTEM DESCRIPTION

Electro-pneumatic precision positioning system. Specification No. 9999. Drawing No. DDDD.

ITEMS TO BE TESTED

Electronic control unit (ECU)

Servo actuator (SA)

Overall system (SYS)

TESTS

(All test specifications will be developed as part of the development test programme.)

ECU

In-circuit test (ICT) on circuit boards. Test specification IIII.

HASS. Test Specification HHHH.

Functional test during HASS. Test Specification FFFF1.

System

Functional test. Test Specification FFFF2.

Final inspection.

PROCEDURES AND RESPONSIBILITIES

All tests are to be performed in accordance with relevant procedures.

REPORTING

All failures are to be reported, analysed and investigated in accordance with Procedure FFFF.

ASSEMBLY AND TEST FLOW

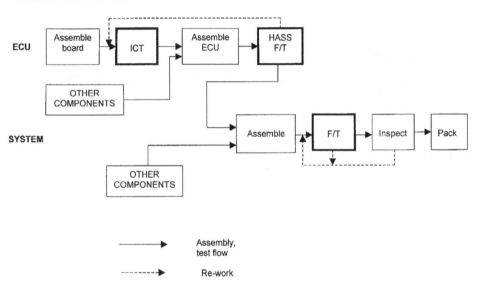

INDEX

Printed and bound in the UK by
CPI Antony Rowe, Eastbourne

Printed and bound by CPI Group (UK) Ltd, Croydon, CR0 4YY

27/10/2024

14580218-0005